THE OZONE DILEMMA

A Reference Handbook

David E. Newton

ABC-CLIO

Santa Barbara, California
Denver, Colorado
Oxford, England

Library of Congress Cataloging-in-Publication Data

Newton, David E.
The ozone dilemma : a reference handbook / David E. Newton.
p. cm.—(Contemporary world issues)
Includes bibliographical references and index.
1. Ozone layer depletion—Handbooks, manuals, etc.
2. Geophysicists—Biography—Handbooks, manuals, etc. I. Title.
II. Series.
QC879.7.N49 1995 551.5'112—dc20 95-18722

ISBN 0-87436-719-0 (alk. paper)

01 00 99 98 97 96 95 10 9 8 7 6 5 4 3 2 1

ABC-CLIO, Inc.
130 Cremona Drive, P.O. Box 1911
Santa Barbara, California 93116-1911

This book is printed on acid-free paper ♾.

Manufactured in the United States of America

THE OZONE DILEMMA

A Reference Handbook

For Bill and Barbara,
who helped us to find our own place in the sun,
for which our many thanks

Contents

Preface

A hole in the sky? Is such a thing possible? How would we ever know if a "hole" were to open up in the atmosphere above us? What difference would it make in our daily lives if such a "hole" were to appear?

People around the world have begun to ask such questions as they learn about the "hole" in the ozone layer that now appears every year above the Antarctic. That "hole" is actually a dramatic loss of ozone in the stratosphere above the regions surrounding the South Pole. The hole was first announced by British scientist James Farman in 1985, although the loss of ozone from the stratosphere was suspected for more than a decade before Farman's own research was published.

Today, scientists are largely in agreement that chemicals invented by scientists and widely used for the past half century by industry are responsible for this loss of ozone. Those chemicals, primarily chlorofluorocarbons (CFCs) and halons, are now being monitored carefully in countries around the world. According to the provisions of the Montreal Protocol on Substances That Deplete the Ozone Layer, a number of those chemicals are now banned or will be phased out of production within the next decade.

The issue of ozone depletion is one fraught with uncertainty and debate, however. Many observers, primarily nonscientists, doubt the connection between so-called ozone-depleting chemicals and the loss of ozone from the stratosphere. Some even doubt that ozone depletion—beyond what normally occurs in the atmosphere—has actually taken place.

The purpose of this book is to outline the main features of this debate and to povide background information on a number of topics related to the issue. Chapter 1 provides a general introduction to the issue of ozone depletion and the role of human-made chemicals in ozone loss. Chapter 2 summarizes some of the most important events that have taken place in this field over the past two centuries. Brief biographical sketches of some important figures in the ozone question are provided in Chapter 3. Chapter 4 contains a number of statements, reports, collections of data, and documents related to the issue of ozone depletion and the chemicals involved. Chapter 5 lists a number of organizations that are interested in this issue. Relevant print and nonprint resources are summarized in Chapters 6 and 7. Finally, there is a glossary of basic terms used in the discussion of ozone depletion, a list of acronyms and abbreviations used in the book, and a listing of chemical formulas and nomenclature for substances involved in ozone depletion.

Ozone Layer Depletion: Myth or Reality?

1

In March 1994, scientists at Oregon State University made a startling announcement. Amphibian populations in many parts of the world were undergoing a noticeable decrease in rates of reproduction, they said. The scientists had discovered the decline in a number of species, ranging from natterjacks in the United Kingdom to salamanders in Mexico. Further, the scientists claimed, this decrease was the result of a thinning of the ozone layer in the Earth's atmosphere. The loss of ozone was allowing higher levels of ultraviolet light to reach the Earth's surface. This ultraviolet light was destroying eggs produced by amphibians, apparently accounting for the reduction in rates of reproduction. (Hileman 1994)

A "Hole" over the Antarctic

The Oregon State announcement was only the latest—if perhaps the most dramatic—of a long series of reports concerning depletion of ozone levels in the Earth's atmosphere. Those reports go back at least to a 1985 discovery by British scientists that the concentration of ozone in the atmosphere over the Antarctic had decreased by as much as 40 percent over the preceding decade. That loss of ozone has since been characterized as a "hole" in the stratosphere's ozone layer.

The appearance of an ozone hole was not just a matter of academic interest. Ozone in the Earth's atmosphere filters out a large fraction of the ultraviolet radiation reaching the planet's upper atmosphere from the sun. Since ultraviolet light has a number of harmful effects on plant and animal life (such as the destruction of amphibian eggs), any reduction in the ozone layer could result in an increase in health problems for living organisms on the Earth's surface.

Studies since 1985 have shown that original reports of an ozone hole were not a one-time phenomenon. The hole has reappeared during every austral (Southern Hemisphere) spring over the Antarctic. In fact, the loss of ozone has increased every year. In addition, evidence of the ozone layer thinning over other parts of the Earth has also begun to appear. A number of scientists, politicians, and ordinary citizens have expressed serious concern about the problem of ozone depletion. They have demanded that governments take action to prevent further damage to the ozone layer.

Not everyone agrees with this assessment, however. Some authorities question the scientific data on which the ozone hole argument is based. Others claim that the problem is not as serious as critics claim. Still others argue that the actions needed to prevent ozone depletion are more expensive than the supposed problem justifies.

This chapter provides a general overview of the ozone depletion issue. The nature of ozone and the ozone layer are described at the outset. Historical trends and current data on ozone depletion are then reviewed. Some theories as to the causes of ozone depletion and its possible effects on living organisms are discussed next. Proposals and actions for reducing ozone depletion are then outlined. Finally, counterarguments about the existence and effects of ozone depletion are presented.

Ozone in the Earth's Atmosphere

Ozone is an allotrope of oxygen. Allotropes are two or more forms of an element with different chemical and physical properties. Molecules of diatomic oxygen, the form of oxygen with which most people are familiar, contain two atoms each, as shown by the formula O_2. In contrast, molecules of ozone contain three atoms each and are represented by the formula O_3.

Ozone is produced by both natural and anthropogenic (human-caused) sources. In the troposphere, the layer of the atmosphere nearest the Earth's surface, ozone is generated in a complex series of reactions associated with the combustion of fossil fuels. It is most commonly found in the air above urban areas where sunlight initiates these reactions among the products released from automobile and truck exhausts. Since tropospheric ozone can cause damage to living organisms, especially plants, it is regarded as a pollutant. Great efforts are now being made to reduce the amount of ozone released to the troposphere.

The situation is very different in the stratosphere, the next higher layer of the atmosphere. There ozone is produced by natural processes and has a beneficial function for terrestrial organisms. In the stratosphere, radiant energy from the sun can cause an oxygen molecule to break apart into two oxygen atoms:

$$O_2 \rightarrow O + O$$

Each oxygen atom can then react with other oxygen molecules to form ozone:

$$O + O_2 \rightarrow O_3$$

The above equation is often written with a double arrow, as shown below, to indicate that, once formed, ozone tends to break down, forming diatomic oxygen (O_2) and monatomic oxygen (O):

$$O + O_2 \leftrightarrow O_3$$

The formation of ozone (the left-to-right equation) occurs largely during daylight hours, and the destruction of ozone (the right-to-left equation) takes place primarily at night.

As the earth's atmosphere evolved, the two opposing reactions described above have reached a state of equilibrium. The rate at which ozone is forming is equal to the rate at which it is breaking down. Today the concentration of ozone in the stratosphere amounts to about ten parts per million (10 ppm). If all that ozone were transported to the Earth's surface, it would fill a layer only 3 millimeters (0.1 inch) thick. In the stratosphere, however, it is spread out in a layer that extends from a height of about 15 kilometers (9 miles) to about 30 kilometers (18 miles) above the Earth's surface.

The concentration of ozone in the stratosphere fluctuates normally from time to time during the day and throughout the year. Changes in the amount of solar radiation reaching the Earth's atmosphere at any one moment are an important factor in accounting for these fluctuations. Still, on an average, the amount of stratospheric ozone on a long-term basis is relatively constant.

The ozone layer is important to life on Earth for two primary reasons. First, like carbon dioxide, it traps some of the heat radiation from the Earth. Thus, it contributes to the greenhouse effect that maintains the planet's temperature at a level that supports life as we know it. Second, ozone molecules in the stratosphere absorb ultraviolet radiation with wavelengths from 240 to 320 nanometers, so-called ultraviolet-B (UVB) radiation.

The latter effect is particularly crucial because UVB radiation has a number of harmful effects on living organisms. It can kill one-celled organisms such as bacteria and algae. It also causes damage to DNA, the molecule that encodes genetic information in all living cells. This damage can result in deformed or diseased plants and animals. Finally, and perhaps most important, UVB radiation can cause skin cancer.

Some Potential Threats to the Ozone Layer

Concerns about damage to the ozone layer go back to the early 1970s. A number of scientists were concerned about two possible threats to the ozone layer: rockets fired off by the U.S. National Aeronautics and Space Administration (NASA) and the proposed supersonic (SST) aircraft. Calculations suggested that chlorine atoms released by rockets and SSTs could bring about the decomposition of ozone molecules with a subsequent deterioration of the ozone layer. This concern was one factor in the U.S. government's decision not to support the development of an SST airplane in the United States. (Cooper 1992)

In 1974, however, an even more troubling threat to the ozone layer was identified. Two American scientists, Mario Molina and F. Sherwood Rowland, hypothesized that a group of compounds known as chlorofluorocarbons (CFCs) could, like rockets and SST aircraft, release chlorine in the stratosphere and damage the ozone layer. They predicted that CFCs could cause a 7 to 13 percent loss of ozone over the next hundred years. (Molina and Rowland 1974)

Chlorofluorocarbons are simple organic compounds consisting of carbon, chlorine, and fluorine. They were first introduced commercially in the 1930s as refrigerants under the name of Freon®. After World War II they became very popular as propellants in aerosol sprays used in a variety of commercial products such as deodorants, hair sprays, spray paints, pesticides, and similar materials.

Specific members of the chlorofluorocarbon family are commonly known by designations such as CFC-11, CFC-12, CFC-112, and CFC-113. These acronyms make it clear to a chemist exactly which chlorofluorocarbon is intended. For example, the correct chemical name for CFC-11 is trichlorofluoromethane, and its chemical formula is CCl_3F. Each individual chlorofluoromethane has properties and uses somewhat different from other members of the family. A list of the most widely produced CFCs, along with their chemical formulas, can be found on page 68.

For many years, scientists and industrial corporations considered CFCs to be the almost ideal commercial product. They are relatively inexpensive to manufacture and are stable, nonflammable, odorless, and nontoxic. Production of CFCs rose from less than 100,000 metric tons in 1950 to more than 800,000 metric tons in 1970. (Cooper 1992)

By the mid-1970s, however, concerns about CFCs in the atmosphere—such as those expressed by Molina and Rowland—began to grow. Research showed that solar energy could cause a CFC molecule to break apart, releasing a free chlorine atom in the process. That chlorine atom could then attack and destroy an ozone molecule. At the conclusion of this reaction, the chlorine atom is regenerated, ready to attack and destroy another ozone molecule.

Molina and Rowland calculated that a single chlorine atom produced from a single CFC molecule could destroy thousands of ozone molecules. In addition, given the stability of CFC molecules, they could be expected to remain in the atmosphere for dozens or hundreds of years. We now know that the lifetime of a single molecule of CFC-11 is 60 years and that of CFC-12, 120 years.

By 1978 the U.S. Environmental Protection Agency (EPA) had become convinced of the threat posed by CFCs to the Earth's atmosphere and banned their use as propellants in aerosol cans. The EPA had less success, however, in restricting CFC use in other applications, such as refrigeration systems, as a cleaning material, and in the manufacture of plastic foam materials. Industries argued that the scientific evidence about damage to the ozone layer was still inconclusive. In addition, they claimed that no other product was available that could be used so effectively in so many important ways. In fact, these claims were essentially correct. Many scientists were still uncertain about the nature of ozone depletion and, in the late 1970s, no alternative compounds were available with which to replace the suspected chlorofluorocarbons.

The Ozone Hole: From Theory to Reality

The debate over depletion of the ozone layer took an unexpected and dramatic turn on May 16, 1985. On that date, a team of British scientists based at Halley Bay, Antarctica, and headed by Dr. J. C. Farman reported in the journal *Nature* on a surprisingly large decrease in the concentration of ozone over the Antarctic. Based on studies going back to 1956, the team announced a loss of ozone far in excess of that predicted by Molina and Rowland. In fact, they had measured a decrease in ozone levels of more than 40 percent in the period between 1977 and 1984. They argued that anthropogenic chemicals—especially CFCs—may have been responsible for this effect. (Farman, Gardiner, and Shanklin 1985)

The Farman report jolted scientists from around the world into action. They soon confirmed the Halley Bay findings and concluded that the "hole" in the ozone layer covered an area of more than 40 million square kilometers (15 million square miles). In 1987, about 150 scientists from 19 organizations and four nations met in Punta Arenas, Chile, to plan the most ambitious study to date of the ozone hole, the Airborne Antarctic Ozone Experiment (AAOE).

Research on the ozone layer like that of the AAOE uses a wide variety of equipment and techniques. Balloons are sent into the stratosphere to capture air samples to determine the substances present there. The AAOE also uses a converted DC-8 passenger plane and a modified version of the U-2 spy plane (renamed the ER-2) to fly through the stratosphere and take air samples.

In addition, satellites are employed to collect data. One of the most useful devices has been the Total Ozone Mapping Spectrometer (TOMS) originally located on board NASA's *Nimbus-7* weather satellite. TOMS has sent back computer-generated "pictures" of ozone concentrations that dramatically show the formation and disappearance of the ozone hole during each austral spring.

By 1987, most nations of the world had become convinced that the Earth's ozone layer was threatened and that CFCs were a major culprit in this problem. They met that year in Montreal to sign the Protocol on Substances That Deplete the Ozone Layer, whereby they agreed to freeze the emission of halons (CFC-like compounds) at 1986 levels by 1992 and to cut CFC emissions in half by 1998. They further agreed to reduce CFC production by 50 percent by 1999. (Benedick 1991)

Additional Discoveries about Ozone Depletion

Meanwhile, research on ozone depletion continued with surprising and troubling results. The hole over the Antarctic reappeared each September and then disappeared a few months later as ozone levels returned to normal. The hole was smaller in 1988 than it had been in previous years, but by 1989 it had grown to its largest size ever. In each succeeding year, the loss of ozone accelerated. In September and October 1993, a TOMS instrument (carried on the Russian *Meteor-3* satellite) detected an ozone concentration of less than 100 dobson units (dobson units are used to measure ozone concentrations). This reading was less than a third of ozone concentrations measured in 1986. During the period 1985 through 1994, scientists occasionally described the status of the ozone layer over the Antarctic as "essentially gone" or "95 percent lost." (Wagner 1994)

At the same time, scientists were detecting evidence that an ozone hole might soon be developing over the Arctic. That evidence was in the form of the discovery of chemicals produced during the breakdown of CFCs and implicated in the decomposition of ozone. A hole over the Arctic would be more troubling than one over the Antarctic. There are more land masses and larger populated areas around the North Pole, so the loss of ozone there is likely to have more effects on plants and animals than is the case at the South Pole. (Chipperfield 1993)

The loss of ozone over nonpolar regions of the planet was also being reported. In 1990, for example, scientists at NASA announced that the ozone layer worldwide had been depleted by 2 to 3 percent over the preceding two decades. Two years later NASA scientists took the unusual step of calling a news conference about the ozone studies even before those studies had been completely analyzed. They announced that the Earth's ozone layer worldwide had been depleted by 2 to 3 percent *in the preceding year*. Ozone loss varied according to latitude, with virtually no loss above the equator and substantially higher loss at higher latitudes. They said that their observations were significantly in excess of what had been predicted by existing theories. (Cooper 1992)

The NASA results were regarded as highly dependable because they had been confirmed by a variety of independent measuring instruments, including the TOMS on the *Nimbus-7* satellite, a second TOMS on the Russian *Meteor-3* satellite, the SBUV/2 instrument in the U.S. National Oceanographic and Atmospheric Administration's *NOAA-11* satellite, the World Standard Dobson Instrument 83, and 22 ground-based measuring instruments. As a result of their studies, the NASA scientists predicted the appearance of a new ozone hole over the Northern Hemisphere soon after the year 2000.

Theories about the Mechanism of Ozone Depletion

The fundamental question facing scientists about depletion of the ozone layer is how the process takes place. Are CFCs in fact the culprit, as many scientists have been arguing? Or are there other mechanisms that contribute to the loss of ozone from the stratosphere?

The CFC theory is probably the most popular explanation among scientists today. According to this theory, CFC molecules that reach the Earth's stratosphere may be broken apart by solar radiation, resulting in the formation of free chlorine atoms.

$$\text{CFC} \xrightarrow{\text{solar radiation}} \text{Cl} + \text{other products}$$

A free chlorine atom can then attack an ozone molecule, producing chlorine monoxide (ClO) and diatomic oxygen (O_2).

$$Cl + O_3 \rightarrow ClO + O_2$$

In a complex series of reactions, chlorine monoxide is then converted to atomic oxygen and more free chlorine atoms.

$$ClO + ClO \rightarrow Cl + Cl + O_2$$

The free chlorine atoms are then available for the destruction of more ozone molecules.

The important point about this sequence of reactions is that once a CFC molecule breaks down to release a free chlorine atom, that chlorine atom can eventually destroy hundreds or thousands of ozone molecules through the repetition of these reactions.

Many scientists are now convinced that these reactions explain the loss of ozone in the stratosphere. They point to measurements that show that regions of the atmosphere high in chlorine monoxide are low in ozone, and vice versa. In fact, the presence of chlorine monoxide is commonly used as a "marker" for the actual or potential loss of ozone. Concerns about ozone depletion over the Arctic and the Northern Hemisphere, for example, are often based on the growing concentration of chlorine monoxide found in the atmosphere above these regions.

Another matter of concern to environmentalists today is a group of compounds containing bromine, especially methyl bromide (CH_3Br) and the halons. Methyl bromide is a naturally occurring compound that is also produced commercially for use as a fumigant. Recent research has shown that grass and forest fires are also a major source of methyl bromide. The halons are similar to CFCs except that they contain the element bromine in addition to carbon, chlorine, and fluorine. Halons are widely used throughout the world in fire extinguishers.

Bromine belongs to the same chemical family as chlorine and, therefore, has similar chemical properties. Scientists now believe that methyl bromide and halons undergo reactions in the atmosphere similar to those for CFCs. In those reactions free bromine atoms (Br) and bromine monoxide (BrO) are formed. Free bromine atoms are then able to "scavenge" on ozone molecules and convert them to diatomic oxygen, just as free chlorine atoms are known to do.

Some scientists question the role of CFCs, methyl bromide, halons, and other synthetic compounds in ozone depletion. They argue that natural processes may be responsible for observed changes over the Antarctic over the last few decades. The key element in this "naturalistic" explanation is the phenomenon of polar stratospheric clouds that form over the Antarctic.

Polar stratospheric clouds form during the austral winter when temperatures may drop as low as –80°C. At those temperatures, many compounds of nitrogen and oxygen, the most abundant elements in the atmosphere, condense and freeze to solid particles. Nitric acid, familiar to most chemistry students, becomes a solid at the very cold temperatures in polar stratospheric clouds.

The significance of these changes is that, at higher temperatures, nitrogen compounds react readily with any free chlorine in the stratosphere, tying it up in the form of compounds such as chlorine nitrate, $ClONO_2$. In this form, the chlorine is not free to react with ozone.

As polar stratospheric clouds develop, then, nitrogen compounds become solid and are no longer able to react with free chlorine. The concentration of free chlorine increases and more ozone is removed from the atmosphere. Voila! The ozone hole is explained on the basis of purely natural processes.

The key element that still remains to be explained in the naturalistic theory is the sudden and dramatic decrease in ozone levels observed since the mid-1970s. One can legitimately argue that sudden changes are not unusual in the natural world. Large increases and decreases in the concentration of stratospheric ozone may, in fact, be quite normal. Given the very short period of time during which such measurements have been carefully made, that explanation is certainly conceivable.

In fact, scientists have given careful attention to one natural factor that may significantly alter the ozone layer: volcanic eruptions. Is it possible, they have asked, that the massive release of dust and gases into the atmosphere from a volcano could produce major changes in the ozone layer? Two eruptions have allowed scientists to test this hypothesis, that of Mount Pinatubo in the Philippine Islands in June 1991 and that of El Chichón in Mexico in April 1982. As a result of these eruptions, huge quantities of sulfur compounds were released to the atmosphere. For example, scientists estimate that the eruption of Pinatubo added about 20 million metric tons of sulfur dioxide to the atmosphere. (McGraw-Hill 1994)

There are two explanations for the relevance of sulfur compounds in the stratosphere to ozone depletion: (1) The presence of such compounds apparently increases the rate at which free chlorine is released from chlorine nitrate and related compounds. An increase of sulfur compounds in the stratosphere, then, could result in a higher concentration of free chlorine and a corresponding increase in the rate of ozone destruction. (2) Sulfur compounds in the atmosphere absorb heat radiated from the Earth. Thus, they contribute to the greenhouse effect resulting primarily from the presence of carbon dioxide in the Earth's atmosphere. The warming effect by sulfur compounds may cause significant changes in atmospheric circulation patterns that could, in turn, result in changes in ozone concentrations over the Antarctic. These changes provide yet another possible explanation for ozone depletion, an explanation that largely ignores any chemical changes involving CFCs, halons, or other synthetic compounds.

The major feature of this physical explanation for ozone depletion is that air masses are constantly in motion throughout the atmosphere. These masses are heated most intensely above the equator, causing them to rise high into the atmosphere. There they are cooled and pushed toward the poles, where they again reach the Earth's surface. They then travel along the surface of the Earth back to the equator, where the whole circulation pattern begins again.

Some scientists argue that changes in these global circulation patterns alone can explain ozone depletion. Factors that we do not yet understand, they suggest, but that might include volcanic eruptions, could dramatically alter those circulation patterns, increasing, decreasing, or displacing the flow of air above the poles. The "loss" of ozone that has been reported, these scientists hint, might not be a destruction of ozone molecules, but a displacement of ozone from one part of the atmosphere to another

If polled, a majority of knowledgeable scientists would probably agree that CFCs, halons, and other synthetic chemicals are implicated in the loss of ozone from the stratosphere. Less agreement is likely, however, as to the exact mechanisms by which ozone depletion takes place and the relative contributions of natural and anthropogenic factors.

Some Potential Effects of Ozone Depletion

Concerns about depletion of the ozone layer are based primarily on the effects that would result from an increase in levels of UVB radiation that would reach the Earth's surface as a result. The next question to be answered, therefore, is a simple and obvious one: Is there evidence of an increase in UVB radiation at the Earth's surface in regions where ozone depletion appears to have occurred? One answer to that question is available from a series of studies conducted in the late 1980s by J. E. Frederick and H. E. Snell at the McMurdo Research Station in Antarctica. The Frederick-Snell team found that UVB radiation increases by as much as 50 percent when the ozone hole appears in comparison with normal ozone concentrations during the austral winter. (World Meteorological Organization 1990)

Similar data for global UVB levels are more difficult to obtain. Instruments used to measure solar radiation reaching the Earth's surface are generally not sufficiently sensitive to the band of UVB radiation blocked out by ozone. Studies that have been done, however, provide a hint as to possible patterns. Various satellite measurements, for example, appear to show an increase in UVB radiation from 1979 to 1988 of 2 to 10 percent between latitudes 30°and 60° south and by 10 to 60 percent under the Antarctic hole. (World Meteorological Organization 1990)

An interesting study showing the influence of the Antarctic hole on UVB levels was published in 1992 by G. Seckmeyer and R. L. McKenzie. These researchers measured the levels of UVB radiation in New Zealand and Germany on identical days in 1990–1991. They found levels to be twice as high in New Zealand as in Germany, an effect they attributed to the reduced levels of ozone over the Southern Hemisphere. (G. Seckmeyer and R. L. McKenzie 1992)

An even more dramatic local effect was reported in 1993 by two scientists at Environment Canada, James Kerr and Thomas McElroy. Kerr and McElroy found that wintertime levels of UVB radiation reaching Toronto had increased by more than 5 percent each year between 1989 and 1993. Kerr and McElroy attributed these increases to downward trends in ozone concentrations measured over Toronto during the same period. (Kerr and McElroy 1993)

Ultraviolet radiation can have either positive or negative effects on living organisms on the Earth's surface. On the positive side, the radiation is necessary for the production of vitamin D, which is essential in the formation of strong teeth and bones. Far more evidence of the harmful effects of UVB radiation has been obtained, however, at every level of terrestrial life. Phytoplankton (tiny plants—usually algae—that live in the water), for example, appear to be especially vulnerable to increases in UVB radiation. Studies of these microorganisms conducted since 1985 suggest that populations have dropped by as much as 12 percent in the seas surrounding the Antarctic. (Stetson 1992) These findings are troubling because phytoplankton are the first link in a complex food web that includes most aquatic organisms in the Southern Hemisphere.

UVB radiation is believed to have a harmful effect on many forms of plant life also, although some questions have been raised about this relationship. In 1989, for example, the United Nations Environment Programme (UNEP) reported that two-thirds of the 300 crops on which data existed are sensitive to increasing levels of UVB radiation. Included among those crops are beans, cabbages, melons, mustard, peas, potatoes, rice, soybeans, sugar beets, and tomatoes. (Caldicott 1992)

Some research raises doubts about the extent of plant damage from UVB radiation, however. Studies of soybeans have shown that in some varieties crop yields decrease with exposure to UVB radiation, but other varieties appear not to be affected by the radiation. (Zurer 1993)

The effects of UVB radiation on humans and other animals tend to fall into three major categories: skin cancer, cataracts and other eye damage, and damage to the immune system. The 1989 UNEP study cited above predicted an increase of 300,000 cases of skin cancer by the year 2000 as a result of ozone depletion that has already occurred. The majority of those cases would appear, UNEP said, in Argentina and Australia because of their proximity to the Antarctic hole. Looking to the future, UNEP warned that a loss of 10 percent of the global ozone layer would result in a 26 percent increase in basal and squamous-cell cancers, two relatively simple forms of cancer. (Caldicott 1992)

The outlook is also not promising for a more deadly form of skin cancer known as melanoma. One group of researchers has estimated that the same 10 percent increase in UVB levels could result in a 9 percent increase in cases of melanoma. (Makhijani, Gurney, and Makhijani 1992) These concerns prompted one medical writer in 1991 to ask "Is Ozone to Blame for Melanoma Upsurge?" (Skolnick 1991)

An increase in cataracts and other eye disorders may also result from higher UVB radiation levels. Early reports from southern Chile claimed that "widespread blindness" had begun to occur among cattle, horses, sheep, and rabbits as a result of ozone depletion over the Antarctic. The 1989 UNEP study predicts that similar effects among humans could results in as many as 1.6 million additional cases of blindness throughout the world each year.

A number of studies have been conducted on the effects of UVB radiation on the immune system. In 1993, for example, Amminikutty Jeevan and Margaret L. Kripke at the University of Texas' M. D. Anderson Cancer Center announced that mice, rats, and guinea pigs exposed to UVB radiation fail to produce immune responses to a variety of allergens. The animals were less resistant to infections and demonstrated inadequate immune responses to microorganisms injected into them. Jeevan and Kripke also reported that similar results were observed in a small sample of human volunteers. Based on this line of research, Kripke had earlier testified before Congress in 1991 that ozone depletion could have a "devastating impact on human health." (*Global Change Research: Ozone Depletion and Its Impacts* 1992)

Some Proposed Solutions for Ozone Depletion

So what is to be done about the problem of ozone depletion? Initial responses to 1985 reports of a hole in the ozone layer ranged from near-hysteria to the absurd. President Ronald Reagan's Secretary of the Interior Donald Hodel argued that the problem was vastly overstated. Even if scientists were correct about ozone depletion, an increase in UVB radiation, and a corresponding rise in skin cancer rates, the government need not be concerned, he said. Since people choose to go outside for a suntan, skin cancer fell into the category of a "self-inflicted disease" that was their own fault, not a public health problem. Hodel recommended that any problems created by ozone depletion could be solved by having people wear broad-brimmed hats, sunglasses, and sunscreen when they go outside. Such a solution, he argued, would be far less expensive than to cut back on CFCs and other critical industrial chemicals. (Cagin and Dray 1993)

The Reagan administration quickly distanced itself from Hodel's remarks, although many officials retained serious doubts about the severity of the ozone problem. Since 1985, a number of more serious solutions to the problem of ozone depletion have been suggested. These solutions fall into three major categories: technological, political, and economic.

Technological Approaches

Technological "fixes" are based on the premise that scientific approaches can be found to counteract an environmental problem. In the case of ozone depletion, one suggestion has been to release simple hydrocarbons such as ethane or propane into the atmosphere. These compounds are known to react with the free chlorine atoms from CFCs that initiate ozone depletion.

Another proposed technological fix is to bombard CFCs in the troposphere with laser beams. This treatment would destroy the CFCs before they get into the atmosphere, thus reducing the damage they do to ozone there.

It is relatively easy to test a technological fix in the laboratory. When such tests are made on the preceding suggestions, both appear to be workable schemes. But other factors are involved in any actual use of the techniques in the real world. In the case of hydrocarbon release, for example, experts estimate that hundreds of aircraft would be needed over an extended period of time in order to carry out the spraying necessary to have any effect on CFCs in the atmosphere.

The laser beam approach also has some major drawbacks. Beams would somehow have to be aimed at CFCs rather selectively, or other beneficial molecules might also be destroyed. In addition, the cost of energy needed to produce the laser beams would be very high, once more probably more than governments would be willing to pay.

The fact that technological fixes such as these are even suggested is an indication of the seriousness with which some scientists regard the problem of ozone depletion. They have been proposed only because those scientists fear that the worst predictions about ozone depletion may, in fact, come true.

Another technological approach to the problem of ozone depletion that has received major consideration is the search for substitutes for CFCs, halons, and methyl bromide. Attention in this research has focused on compounds that are similar to ozone-depleters, but that have one or more hydrogen atoms in their molecules. These compounds are known as hydrochlorofluorocarbons (HCFCs) and hydrofluorocarbons (HFCs). (At the present, no satisfactory substitute for methyl bromide has been found.)

The advantage of HCFCs and HFCs is that they are similar in many ways to CFCs and halons, but they tend to break down more quickly in the atmosphere. The atmospheric lifetime for the hydrogen-containing substitutes ranges in most cases from less than a year to about five years. (World Meteorological Organization 1990)

Still, HCFCs and HFCs are by no means perfect replacements for their now-popular cousins. Both groups of compounds are decomposed by solar radiation, just as are CFCs and halons. In fact, the HCFCs and HFCs are also regulated by the Montreal Protocol and are scheduled to be totally phased out in the year

2030. Between 2015 and 2030, nations are encouraged to begin reducing their dependence on these compounds, as they were to have done in the late 1990s with CFCs and halons.

Even these dates are regarded as too far into the future by some environmental scientists. Scientists at the Institute for Energy and Environmental Research, for example, claim that the ozone-depleting effects of HCFCs and HFCs have been drastically underestimated and that they should be phased out long before 2030.

Political Approaches

Nearly everyone agrees that the best way to deal with the problem of ozone depletion is through prevention. That approach relies on the reduction or elimination of CFCs, methyl bromide, halons, and other harmful chemicals released into the atmosphere. The U.S. government acknowledged the value of this approach more than a decade ago, even before the hole over the Antarctic was discovered. It took a leadership role in organizing the Vienna Convention for the Protection of the Ozone Layer, convened under the auspices of the United Nations Environmental Programme.

That convention came to a successful conclusion in March 1985 with the signing of an agreement by 20 nations. The Vienna agreement called for an aggressive program of research on ozone depletion and chemicals that may be responsible for that phenomenon. It also outlined an agenda of meetings, the end result of which was to be an international agreement on the banning of ozone-depleting compounds.

That final step took place on September 16, 1987, in Montreal. After nine months of negotiations, 24 nations signed the Montreal Protocol on Substances That Deplete the Ozone Layer. Although the number of nations that signed the protocol was relatively small, those nations are responsible for 99 percent of the production and 90 percent of the consumption of CFCs in the world.

The primary feature of the Montreal protocol was an agreement to phase out the use of ozone-depleting chemicals according to a rigorous schedule. Specially, signers agreed to

1. freeze the emission of halons at 1986 levels by 1992;
2. freeze the emission of CFCs at 1986 levels by 1990;
3. cut CFC emissions by 20 percent of 1986 levels by 1994;
4. cut CFC emissions by 50 percent of 1986 levels by 1999;
5. cut CFC production by 50 percent by 1999.

In addition, protocol signatories agreed to meet on a regular basis to review new data on ozone depletion and to modify the Montreal Protocol as necessitated by these new findings.

One of the first follow-up meetings, held in 1990, attacked a thorny issue that had not been resolved in Montreal. Developing nations have traditionally viewed global environmental problems, such as ozone depletion and global warming, in a somewhat different light than have developed nations. While recognizing the possible seriousness of such issues, developing nations have argued that they cannot pay the price of making the same industrial cutbacks as developed nations. Their position has been that it is unfair for the nations of the world who have achieved a high standard of living to prevent other nations from achieving similar goals. Recognizing the legitimacy of that argument, industrialized nations established a $240 million fund in 1990 for the development of modern technologies for use in developing nations, technologies that do not make use of CFCs and other ozone-depleting chemicals.

Partly as a result of this action, the fourth follow-up meeting at Copenhagen in 1992 was attended by more than 100 nations. At that meeting, evidence was presented that ozone depletion was taking place even more rapidly than had been suspected in 1985 at Vienna and in 1987 at Montreal. As a result, participants voted to accelerate deadlines for phasing out ozone-depleting chemicals. January 1, 1994, was set as the date on which all halon production was to be halted, and January 1, 1996, was set as the comparable cut-off date for CFCs.

In the meanwhile, the United States, the world's largest producer and consumer of ozone-depleting chemicals, had decided to move forward even more aggressively. The Clean Air Act Amendments of 1990 authorized government agencies to reduce or ban the production and/or use of a number of ozone-deleting chemicals. Acting on that authority, the Environmental Protection Agency (EPA) issued a rule in early 1993 banning the sale of certain products containing CFCs after February 16, 1993, and limiting the sale of such materials to essential products (such as medical sprays) after January 17, 1994. The EPA also announced that it planned a total ban on the production and importation of methyl bromide by the year 2000.

Actions such as those called for by the Montreal Protocol or proposed by the EPA are commonly met with strong opposition. For example, the U.S. Department of Agriculture, the chemical industry, and representatives of agriculture all objected strenuously to the proposed methyl bromide ban. These groups argued that the compound is essential for the safe storage of many kinds of foods and crops and that no satisfactory alternative has yet been developed. They have also questioned some of the scientific data on which the EPA based its decision. Given the apparent widespread acceptance of the present theory of ozone depletion and the development of replacements for ozone-depleting compounds, it appears likely that actions like those called for by the Montreal Protocol and the EPA will continue to go forward.

Economic Approaches

Another approach to solving the ozone depletion problem is to make use of economic incentives. An example of this approach is the four-pronged program for limiting the release of CFCs into the atmosphere by U.S. corporations announced by the EPA in 1992. The first element of this program involves a marketable permits system. Under this system, each of the seven U.S. companies that produces CFCs or halons is permitted to produce a certain limited amount of these chemicals. But each company can distribute the amount of individual chemicals produced to meet market needs, as long as their total production falls within the EPA-assigned limits. Thus, a company might make a relatively large quantity of CFC-113 and a relatively small amount of CFC-115 in one year and then reverse that pattern in another year. Companies can also exchange allowances with each other so that more efficient companies can produce the CFCs for which their plants are most suitably designed.

A second element of the EPA plan involves an excise tax on ozone-depleting chemicals. The EPA has said that it believes that the additional cost of selling ozone-depleting chemicals caused by the excise tax is one reason that sales of these chemicals dropped in the first two years its plan was in operation. (Lee 1992)

A third component of the EPA plan focuses on guaranteeing that only safe alternatives are used as replacements for CFCs, halons, and other chemicals that have been or will be banned. Specifically, the agency now reviews all alternatives for potential health and environmental effects, including ozone depletion, toxicity, worker and consumer safety, and global warming potential.

Finally, the EPA program includes a national recycling component. Regulations have been established that require companies to reuse ozone-depleting materials rather than release them into the atmosphere. For example, refrigerants used in an outmoded refrigeration system must be removed and then reused in newly manufactured equipment.

Differences of Opinion about Ozone Depletion

The case for ozone depletion as presented thus far may seem like a relatively straightforward story. Such is not the case. Every aspect of this story—from the extent of ozone loss to the causes of any such losses to potential solutions for ozone depletion—is the subject of intense debate. Representatives of industry, for example, have long expressed doubts about the reality of problems with the ozone layer. In an argument familiar to those who doubt the seriousness of many environmental problems, they have tended to claim that sufficient scientific data are not yet available on which to conclude that ozone depletion has occurred. As a result, they recommend that little or no action be taken to deal with a problem that may not even exist.

In February 1992, for example, five years after the signing of the Montreal Protocol, an editorial written for the strongly business-oriented *Wall Street Journal* responded to NASA's unusual news conference of the preceding week. The writer observed that NASA's data were still preliminary and "based on the flight of one converted ER-2 airplane." "Arguing the destruction of the ozone layer on the basis of one day's, or a few weeks', data," the writer continued, "is a bit like announcing the comeback of retail sales on evidence that takes Neiman-Marcus' receipts from the day before Christmas and annualizes them." (*Wall Street Journal* 1992)

In response, a NASA scientist involved in the research criticized by the *Wall Street Journal* writer noted that the writer has simply ignored large volumes of data that had been collected over the preceding decade. (Anderson 1992)

The 11 June 1993 issue of *Science* discussed in detail the "backlash" that has developed against the case for ozone depletion. Writer Gary Taubes pointed out in that article that the number of scientists involved in this "backlash" is relatively small and tends to include many who are not specialists in atmospheric sciences. As one example, zoologist Dixy Lee Ray, formerly chair of the Atomic Energy Commission and formerly governor of the state of Washington, has called the ozone depletion theory "balderdash" and "poppycock."

One of the authorities on whom Ray relied for much of her scientific information was S. Fred Singer, former chief scientist at the U.S. Department of Transportation. Singer has written and spoken on a number of occasions about the undue emphasis scientists have placed on ozone depletion. In an October 1991 article in *Consumers' Research* magazine, for example, he argued that global environmental issues have become "the plaything of international lawyers" who become "very upset" when scientific evidence does not support their "global environmental scams."

Critiques by individuals such as Ray and Singer have also been taken up by political conservatives who oppose spending money on environmental issues they do not believe exist. Among these has been political commentator Rush Limbaugh, who has called the case for ozone depletion a "scam" put forward by government scientists to secure their jobs and maintain or increase funding for their research.

A common flaw in the presentations of Ray, Singer, Limbaugh, and other critics of the ozone issue is their tendency to ignore large bodies of data and/or to misrepresent fundamental scientific information. For example, the June 1993 issue of *Omni*, a quasi-scientific magazine, carried an article by novelist James P. Hogan. Hogan described the ozone hole phenomenon as an example of "environmentalist mania," for which there was essentially no scientific evidence. In a sidebar to the article, physicist Frederick Pohl responded to Hogan's claims. Pohl explained that one problem with Hogan's presentation was that he (Hogan) simply had some of his science wrong. As one example, the fact that CFCs are more dense than other atmospheric gases does not mean, as Hogan suggests, that they are physically incapable of diffusing into the stratosphere.

Deciding what to do about ozone depletion, to the extent that it has occurred, is another point of dispute. Deterioration of the ozone layer, like the global warming that many people believe has begun, is a new kind of environmental problem. One cannot see obvious unpleasant changes around them, as is usually the case with air pollution, water pollution, and other types of environmental degradation. Loss of stratospheric ozone takes place in distant parts of the atmosphere at a relatively slow rate. Any health or environmental problems it produces may not show up for years or decades.

The issue in such cases is whether to take actions now that may be very expensive and cost jobs to rectify a problem so inconspicuous and seemingly so remote from daily life. As Pohl commented in his *Omni* sidebar, the question is "If the consensus of most scientists is wrong, and there is, after all, no danger to the ozone layer, then doing what this consensus suggests will unnecessarily cost us all some money and inconvenience. But if the scientists are right and we do nothing, it will cost us a great deal more money, a great deal more inconvenience, and a very great deal of suffering and human lives." (Pohl 1993)

The opposition of industries to policies such as those outlined in the Montreal Protocol and the Clean Air Act Amendments of 1990 are, therefore, easy to understand. Companies can point to specific dollar and job losses that will occur if and when they are required to abandon the use of commercially vital compounds like the CFCs.

During hearings before the House of Representatives Subcommittee on Health and the Environment of the Committee on Energy and Commerce, for example, Elwood P. Blanchard, vice president for the Chemicals and Pigments Department of E. I. du Pont de Nemours & Company, the world's largest producer of CFCs, argued that "we believe there is no imminent crisis that demands unilateral regulation [of CFCs]." (*Ozone Layer Depletion* 1987)

At the same hearing, a policy statement from the Chemical Manufacturers Association's Fluorocarbon Program Panel was read into the record. One conclusion reached by the panel was that "current model calculations, incorporating the best available scientific information, indicate that there is no imminent hazard from continued emissions of CFCs at or near today's rates. Thus, there is no scientific justification for a near-term phase-out or reduction in CFC emissions below today's production levels."

Yet, many industry leaders recognized the potential threat of CFCs and other chemicals to the ozone layer early on and acknowledged that some regulatory actions might be needed. Remarkably, the Du Pont Corporation itself took the lead in acting on this belief. On March 24, 1988, the company announced that it had accepted scientific evidence about the connection between CFCs and ozone depletion. As a result, Du Pont would begin phasing out the production of ozone-depleting chemicals, a move that placed it two years ahead of U.S. government action.

The controversy over ozone depletion became even more complex in late 1994 when a team of researchers reported that the concentration of ozone over North America had rebounded from its lows of the previous two years. The researchers had measured an increase of 15 percent in ozone concentration over the levels recorded in the winter of 1992–1993.

Scientists warned about misinterpreting these data, however. They pointed out that the increase involved a comparison with the lowest ozone readings ever made and that, even with the increase, the 1993–1994 values were far less than normal. In fact, researchers at the University of Athens in Greece reported in the same journal that the level of ozone had decreased by 2.5 percent in the summer and 7 percent in the winter over the previous decade. Obviously, changes are still taking place in the ozone layer.

Conclusion

The debate concerning ozone depletion is far from over. Global processes such as the build-up and destruction of ozone take place in irregular patterns over many years and decades. The contribution of human activities to such processes may be difficult to delineate with a high degree of certainty. At a time when both national and international economic conditions are weak, spending large amounts of money to reduce ozone depletion may be a hard case to make. Citizens of the year 2000 may well be rehashing many of the same issues about ozone depletion that confront the world today.

References

Anderson, James G. 1992. "Letter to the Editor." *Wall Street Journal* (March 23).

Benedick, Richard Elliot. 1991. *Ozone Diplomacy*. Cambridge, MA: Harvard University Press.

Cagin, Seth, and Philip Dray. 1993. *Between Earth and Sky: How CFCs Changed Our World and Endangered the Ozone Layer*. New York: Pantheon Books.

Caldicott, Helen. 1992. *If You Love This Planet: A Plan To Heal the Earth*. New York: W. W. Norton.

Chipperfield, Martyn. 1993. "Satellite Maps Ozone Destroyer." *Nature* (April 15).

Cooper, Mary H. 1992. "Ozone Depletion." *CQ Researcher* (April 3).

Farman, J. C., B. G. Gardiner, and J. D. Shanklin. 1985. "Large Losses of Total Ozone in Antarctica Reveal Seasonal ClO_x/NO_x Interaction." *Nature* (May 16).

Global Change Research: Ozone Depletion and Its Impacts. 1992. Hearings before the U.S. Senate, Committee on Commerce, Science, and

Transportation, November 15, 1991. Washington, DC: U.S. Government Printing Office.

Hileman, Bette. 1994. "Amphibian Population Loss Tied to Ozone Thinning." *Chemical & Engineering News* (March 7).

Kerr, J. B., and C. T. McElroy. 1993. "Evidence for Large Upward Trends of Ultraviolet-B Radiation Linked to Ozone Depletion." *Science* (November 12).

Lee, David. 1992. "Ozone Loss: Modern Tools for a Modern Problem." *EPA Journal* (May–June).

Makhijani, Arjun, Kevin Gurney, and Anni Makhijani. 1992. *Saving Our Skins: The Causes and Consequences of Ozone Layer Depletion and Policies for Its Restoration and Protection.* Washington: Institute for Energy and Environmental Research.

McGraw-Hill Yearbook of Science & Technology. 1994. New York: McGraw-Hill, Inc.

Molina, Mario, and F. Sherwood Rowland. 1974. "Stratospheric Sink for Chlorofluoromethanes: Chlorine Atom-Catalyzed Destruction of Ozone." *Nature* (June 28).

Ozone Layer Depletion. 1987. Hearings before the U.S. House of Representatives, Committee on Energy and Commerce, Subcommittee on Health and the Environment, March 9, 1987. Washington, DC: U.S. Government Printing Office.

"Ozone Layer Rebounds, but Caution Still Urged." 1994. *San Francisco Chronicle* (August 27).

Pohl, Frederik. 1993. "Ozone Realities." *Omni* (June).

"Press-Release Ozone Hole." 1992. *Wall Street Journal* (February 28).

Seckmeyer, G., and R. L. McKenzie. 1992. "Increased Ultraviolet Radiation in New Zealand (45° S) Relative to Germany (48° N)." *Nature* (September 10).

Skolnick, Andrew A. 1991. "Is Ozone Loss to Blame for Melanoma Upsurge?" *JAMA* (June 26).

Stetson, Marnie. 1992. "Saving Nature's Sunscreen." *Worldwatch* (March–April).

Wagner, Richard. 1994. "The Ozone, Lost Again." *Ad Astra* (January–February).

World Meteorological Organization. 1990. *Scientific Assessment of Stratospheric Ozone: 1989*. Geneva: World Meteorological Organization.

Zurer, Pamela S. 1993. "Ozone Depletion's Recurring Surprises Challenge Atmospheric Scientists." *Chemical & Engineering News* (May 24).

Chronology 2

1801 German physicist Johann Wilhelm Ritter first recognizes the existence of radiation beyond the violet end of the visible spectrum, radiation now known as ultraviolet radiation.

1840 Ozone is discovered by the Swiss chemist Christian Schönbein, who names it for the Greek word *ozein*, meaning "to smell."

1881 Irish chemist W. N. Hartley hypothesizes that ozone exists in relatively high concentrations in the stratosphere and is responsible for the absorption of ultraviolet light from solar radiation.

1882–1883 During the First Polar Year, scientists from 12 nations work out of 13 research centers in the Arctic, the Antarctic, and Greenland to conduct a massive survey of geological, meteorological, and geophysical conditions in the vicinity of the poles. The research yields fundamental data about atmospheric conditions above the poles.

1899 French meteorologist Léon Philippe Teisserenc de Bort discovers a totally unexpected temperature inversion in the atmosphere (later found to be the lower boundary of the stratosphere). He later suggests the names *troposphere* and *stratosphere* for the two lower layers of the atmosphere.

1913 French physicist Charles Fabry uses highly efficient optical equipment that he has developed himself to prove that high concentrations of ozone exist in the upper atmosphere.

1926 English physicist Gordon M. B. Dobson develops a prism spectrophotometer for making precise measurements of ozone concentrations in the stratosphere. He sets up a worldwide network of observing stations that use his instrument.

1926–1927 English physicist F. J. W. Whipple measures the lower boundary of the stratosphere.

1928 On 31 December, the Frigidaire Corporation is granted a patent for the manufacture of chlorofluorocarbons. The compounds were originally invented by one of its employees, Thomas Midgley, Jr.

1931 English geophysicist Sydney Chapman proposes a mechanism by which ozone is created in the stratosphere. He hypothesizes that solar radiation causes oxygen molecules to break apart and form individual oxygen atoms which, in turn, combine with other oxygen molecules to form ozone.

1932 CFCs are first used in home air-conditioning units, the "Atmospheric Cabinets" produced by Carrier Engineering Corporation.

The Second Polar Year is held, but data collected is never properly processed because of the worldwide Depression.

1941 CFCs are first used as propellants in aerosol pesticides. The invention is soon put to use by the U.S. armed forces in military campaigns in the Pacific area.

1945 The first commercial "bug bombs" for dispensing pesticides with CFCs as propellants are put on sale in New York City.

1957–1958 Sixty-six nations take part in the 18-month long International Geophysical Year (IGY), intended as a follow-up to the First and Second Polar Year. IGY results in the most complete accumulation of data about the planet in history. Among the findings reported during this period is that the ozone layer is significantly thinner than scientists had expected it to be.

1962 Rachel Carson's *Silent Spring* raises questions about the uninhibited use of pesticides, including those dispensed from CFC-propelled aerosol containers.

1963 The U.S. Federal Aviation Agency recommends that the U.S. government actively support the development of a supersonic (SST) aircraft. The recommendation sets in motion an intensive debate about the potential threat to the ozone layer posed by such aircraft. On June 5, 1963, President John F. Kennedy officially asks Congress to appropriate $60 million to begin research on an SST aircraft.

The U.S. Congress passes and President John F. Kennedy signs the Clean Air Act.

1964 Canadian chemist John Hampson suggests that oxides of hydrogen may react with ozone in the stratosphere, causing the gas's concentration to be less than theory had predicted.

1970 Dutch meteorologist Paul Crutzen proposes a mechanism by which oxides of nitrogen are involved in maintaining the concentration of ozone in the stratosphere.

The 1963 Clean Air Act is updated and expanded.

Concerns about environmental issues reach a peak during celebration of the first Earth Day on April 22. Questions about the use of CFCs as aerosol propellants take an important place in these celebrations.

1971 Physicist James McDonald and chemist Harold Johnston suggest that gases produced by supersonic aircraft may contain compounds that can damage the ozone layer. At least partly as a result of this research, the U.S. Congress later votes to discontinue funding for SST research.

The newly created Environmental Protection Agency (EPA) begins operation on December 2.

1971–1974 More than a thousand scientists from ten nations take part in the Climatic Impact Assessment Program (CIAP), an effort to collect extensive quantities of data about the stratosphere.

1972 The United Nations Environmental Programme (UNEP) is created at the U.N. Conference on the Human Environment in Stockholm. UNEP headquarters are set up in Nairobi, Kenya.

1973 University of Michigan researchers Richard Stolarski and Ralph Cicerone report that exhaust gasses from the space shuttle may include chlorine. They warn that shuttles may, therefore, cause damage to the ozone layer. NASA scientists are not convinced, however, that this problem is serious enough to interrupt planning for future shuttle launches. The journal *Science* rejects the Stolarski-Cicerone paper partly because one reviewer claims that it "is of no possible geophysical significance."

1974 On June 28, F. Sherwood Rowland and Mario Molina's historic paper on the potential effects of CFCs on the ozone layer is published in *Nature*.

Department of Commerce statistics indicate that the production of CFCs in the United States involves about 600,000 jobs with a total payroll of $6.7 billion. Total value of all CFC products produced is estimated at $500 million.

Committees of the U.S. Congress hold initial meetings on ozone depletion.

1975 The National Resources Defense Council files suit demanding that the U.S. Consumer Product Safety

1975 *cont.* Commission ban the use of CFCs in aerosol spray cans. At virtually the same time, the Johnson Wax Company announces that it will stop using CFCs in its products. Johnson is the nation's fifth largest manufacturer of aerosol products.

Oregon becomes the first state to ban CFCs in aerosol products. The Oregon law eventually takes full effect on March 1, 1977.

The Committee on the Inadvertent Modification of the Stratosphere (IMOS) is formed to study the problem of ozone depletion. It reports in June that the issue was "a legitimate cause for concern" and suggests that the National Academy of Sciences study the problem in more detail.

NASA researcher Veerabhadran Ramanathan finds that CFCs are greenhouse gases similar to carbon dioxide and methane. This discovery adds a new dimension to the question of banning the use of CFCs and other ozone-depleting chemicals.

1976 The National Academy of Sciences publishes its first of a number of reports on ozone depletion. The report largely substantiates the hypothesis that CFCs damage ozone in the atmosphere and estimates that continued use of CFCs at 1973 rates will result in an ozone loss of 2 to 20 percent, with 7 percent being the most likely value. The report also calls for additional research on the issue over the next two years.

The U.S. Food and Drug Administration (USFDA) announces its intention to begin a phaseout of all nonessential use of CFCs.

The Governing Council of the United Nations Environment Programme calls for an international meeting to discuss ozone depletion.

1977 James G. Anderson finds a chlorine oxide concentration of 8 parts per billion in the stratosphere above Texas, a level far in excess of that predicted by scientists. The

1977 *cont.* presence of chlorine oxide is accepted as a strong indicator of potential ozone loss in the atmosphere.

A joint committee of the USFDA, EPA, and the Consumer Product Safety Commission announces a schedule for the phaseout of CFCs in nonessential aerosol products.

UNEP sponsors the first international conference on CFCs in Washington, D.C. The meeting is attended by 33 nations and representatives of the European Economic Community. The meeting recommends a "World Plan of Action on the Ozone Layer" and creates a Coordinating Committee on the Ozone Layer (CCOL). CCOL's assignment is to prepare annual reviews of the scientific status of the ozone layer.

A film is released entitled *Day of the Animals*. In the film, animals go berserk and run wild because of increases in ultraviolet light caused by a breakdown of the ozone layer.

1978 The manufacture of all CFC propellants in the United States is banned as of October 21. Companies are allowed to use existing supplies of such materials until December 15.

Canada and Sweden join the United States in banning the use of CFCs in nonessential aerosols.

New York becomes the second state to pass CFC legislation. Manufacturers of aerosol sprays using the compounds are required to include a notice on labels that the products may harm the environment.

A total ozone mapping spectrometer (TOMS) is sent into space on board the *Nimbus 7* satellite. It records ozone concentrations far below those predicted by existing atmospheric theories, but these data are ignored by scientists who assume the instrument is malfunctioning. Analysis some years later showed that the data were, in fact, correct and were some of the earliest records of ozone depletion available to scientists.

1978 *cont.* The Federal Republic of Germany hosts the second international conference on regulation of CFCs.

1979 Interstate shipment of all nonessential aerosol products is banned as of April 15.

The National Academy of Sciences (NAS) releases its second report on ozone depletion. The report predicts a 16.5 percent loss of ozone by the end of the twenty-first century if production and consumption of CFCs continues at its current rate. The NAS panel recommends that the United States take the lead in obtaining international agreement to control the use of CFCs.

The joint committee of the USFDA, EPA, and the Consumer Product Safety Commission decide to delay implementation of Phase Two regulations for limitations on CFC uses. The committee claims that there is too much scientific uncertainty to justify additional regulations at this point in time.

1980 At a UNEP-sponsored meeting in Oslo, the United States, Canada, Denmark, Norway, Sweden, the Netherlands, and the Federal Republic of Germany call for immediate reductions in the use of CFCs. They say that waiting for additional scientific proof of ozone depletion is dangerous.

Ronald Reagan is elected President of the United States. As part of his campaign for the presidency, Reagan has called for a reduction in government regulatory activities. That philosophy translates into a less aggressive approach to regulating ozone-depleting chemicals in the first half of the 1980s.

The European Economic Community announces a 30 percent reduction in the use of aerosols.

1981 UNEP creates an Ad Hoc Working Group of legal and technical experts to develop an international convention for the protection of the ozone layer. The Ad Hoc group holds seven meetings over the next four years in preparation for the Vienna Convention of 1985.

1981 *cont.* At her confirmation hearings, EPA administrator-to-be Anne M. Gorsuch testifies that there is not yet enough scientific evidence about ozone depletion to justify any type of governmental action.

1982 In an action typical of the environmental policies of President Ronald Reagan's administration, his administrator of the EPA, Anne M. Gorsuch, announces that the United States will reduce its contribution to the UNEP budget from $10 million to $7.85 million for the coming year. Reagan's original budget proposal had been to discontinue all funding to UNEP, but Congress had insisted on the $7.85 million level.

A third National Academy of Sciences report on ozone depletion reduces the estimated damage to the ozone layer to 5 to 9 percent, about half that of the second report in 1979.

Mexican volcano El Chichón erupts in April, emitting a variety of chemicals that are later to contribute to depletion of the ozone layer.

An Ad Hoc Working Group of Legal and Technical Experts for the Preparation of a Global Framework Convention for the Protection of the Ozone Layer is convened in Stockholm under the auspices of UNEP.

1983 The National Academy of Sciences publishes its fourth report on ozone depletion. The report predicts ozone depletion levels of only 2 to 4 percent, the lowest prediction of any made by an NAS panel thus far.

Anne Gorsuch Burford resigns as administrator of the EPA and is replaced by William Ruckelshaus. Under Ruckelshaus, the EPA once again begins to pursue plans for extending the ban on CFC use, a course of action that had been put on hold under Burford.

Finland, Norway, and Sweden join to write a draft protocol calling for a worldwide ban on the use of aerosols and for strict controls on the production and use of CFCs. The draft protocol eventually becomes known as the Nordic Annex.

1984 F. Sherwood Rowland presents calculations that suggest that the presence of chlorine nitrate in the atmosphere may significantly increase the rate at which ozone is being depleted.

The National Resources Defense Council (NRDC) files suit in Federal District Court in Washington, D.C., asking the EPA to issue rules limiting further emissions of CFCs into the atmosphere. The suit is finally resolved in May 1986 when the NRDC, the EPA, the Alliance for Responsible CFC Policy, and other interested groups agree to a policy that leads to a "multi-faceted work plan of study, scientific review, document development, and regulatory analysis."

1985 Lee Thomas is appointed EPA administrator, replacing William Ruckelshaus.

The Vienna Convention to Protect the Ozone Layer is signed on March 22. The agreement calls for an intensive program of research and information exchange among international scientists on the question of ozone depletion.

James Farman's crucial paper on ozone depletion over the Antarctic is published in the May 16 issue of the journal *Nature*. About a month later, NASA scientist Donald Heath is able to find evidence from earlier *Nimbus 7* flights that vividly illustrate the presence of an ozone hole going back as far as 1979.

UNEP and NASA collaborate in the publication of a massive report on the status of ozone depletion. The report says that damage to the ozone layer has already begun, and scientists know little about future changes that may occur or what effect those changes will have for life on Earth.

1986 NASA conducts the National Ozone Expedition, designed to study ozone concentrations above the Antarctic. The agency follows up a year later with a second Antarctic ozone project, the Antarctic Airborne Ozone Experiment.

1986 *cont.* A team of NASA research scientists led by Arlin Krueger and Richard Stolarski report in the journal *Nature* on ozone concentration readings over the Antarctic extending back to 1978. Those readings confirm the depletion of ozone reported a year earlier by James Farman.

Initial hearings on the ozone depletion issue are held before the Subcommittee on Environmental Pollution of the Committee on Environment and Public Works of the U.S. Senate.

In a dramatic turnaround from its earlier position, the Du Pont Corporation announces that it will support limitations on the future production and use of CFCs.

The journal *Geophysical Research Letters* publishes a special supplement to its November issue in which a group of researchers present evidence for a physical, rather than a chemical, explanation of ozone depletion over the Antarctic.

A report prepared by the Rand Corporation says that CFC emissions in 1986 had reached levels at least as high as those prior to the mid-1970s bans on their use.

NASA creates the 100-plus member Ozone Trends Panel and charges the group with providing accurate, up-to-date, and complete information on the status of stratospheric ozone. The group analyzes all existing data on the topic and announces that stratospheric ozone levels have, indeed, declined significantly between 1978 and 1985.

For the first time, the industry group Alliance for Responsible CFC Policy announces that it will support limits on the growth of CFC production and use.

The U.S. Senate ratifies the Vienna Convention.

The United States and the Soviet Union agree to a joint program of research on stratospheric ozone.

1987 The U.S. Department of the Interior announces a "Personal Protection Plan" for dealing with the threat of ozone

1987 *cont.* depletion. The plan calls for citizens to wear large hats, sunglasses, and suntan lotion in order to avoid the harmful effects of increased UVB radiation.

NASA sponsors the Airborne Antarctic Ozone Expedition to study ozone loss above the South Pole.

The McDonald's Corporation announces that it will no longer use polyurethane containers made with CFCs at its fast-food restaurants.

The Montreal Protocol is signed.

A photograph taken by the TOMS aboard the *Nimbus 7* satellite shows an ozone hole above the Antarctic larger in dimensions than the continent itself.

1988 On January 1, Kenneth P. Bowman of the University of Illinois announces that global ozone levels have decreased by an average of 5 percent between 1979 and 1986.

On March 14, the U.S. Senate votes 83 to 0 to ratify the Montreal Protocol.

The Ozone Trends Panel officially announces for the first time a significant decrease in the concentration of stratospheric ozone over the Northern Hemisphere.

Scientists report that the ozone hole over the Antarctic is smaller than at any time in the preceding decade. The good news lasts only one year, however, before ozone loss begins to increase once more in 1989. In later years, 1988 is sometimes referred to as the "good year."

A trade fair on substitutes for CFCs is held in Washington, DC.

1989 At a March meeting in Brussels, the European Economic Community votes to completely phase out use of CFCs by the year 2000. Almost simultaneously, President George Bush announces that the United States will follow suit if acceptable alternatives to CFCs are available by that date.

1989 *cont.* Airborne Arctic Stratospheric Expedition I (AASEI) attempts to measure ozone losses above the North Pole. Two years later, an extension of that research (AASEII) continues from 1991 to 1992.

1990 The 1963 Clean Air Act is updated and expanded to include provisions for dealing with ozone depletion.

Industrialized nations establish a $240 million fund to support the development of alternatives to CFCs appropriate in Third World nations.

E. I. du Pont de Nemours & Company, the world's largest producer of CFCs, announces that it will eliminate the production of these chemicals by the year 2000. Three years later, in 1993, it promises to speed up this timetable by three to five years.

Representatives of the world's leading CFC producing nations meet in London to sign phase two of the Montreal Protocol. According to the new agreement, production of CFCs will be phased out completely by the year 2000.

Dave Hofmann reports that his research team has found no ozone whatsoever in a layer of the stratosphere between 15 and 17 kilometers above the South Pole.

1991 The eruption of Mount Pinatubo in the Philippine Islands releases large volumes of materials that, scientists believe, are likely to have a dramatic effect on the Earth's ozone layer.

Scientists at the National Aeronautics and Space Administration announce that ozone losses over the United States have amounted to 4 to 5 percent since 1978. Losses for nations at higher latitudes, such as Sweden and Canada, are estimated at 8 percent.

1992 NASA scientists announce that worldwide global levels of ozone have reached the lowest levels ever recorded, about 2 to 3 percent less than 1991 levels. They conclude that an ozone hole over the Arctic may soon appear.

1992 *cont.* Meeting in Copenhagen, delegates from 87 nations agree to amend the Montreal Protocol, ending production of halons on January 1, 1994, and CFCs on January 1, 1996.

Scientists at the genetic engineering corporation Envirogen discover a bacterium that will break down hydrochlorofluorocarbons and hydrofluorocarbons before they can have an effect on ozone.

Experiments on heterogeneous chemical reactions in the stratosphere above the Northern Hemisphere are carried out between October 1992 and May 1993 during the Stratospheric Photochemical Aerosol and Dynamics Experiment (SPADE).

Member nations of the European Economic Community announce that they will end CFC production as of January 1, 1995.

The U.S. Senate votes unanimously to end production of CFCs in the United States by 1996. In response to the Senate's action, President George Bush announces an aggressive program by the United States government to cut back on the production and use of CFCs. He says that all production of CFCs except for that needed in indispensable materials will be ended effective December 31, 1995.

1993 An EPA rule banning the sale of certain products containing CFCs takes effect on February 16.

Representatives from more than 100 nations meet in Copenhagen for the fourth follow-up meeting of the Montreal Protocol.

Scientists at NASA announce a dramatic decrease in ozone concentrations between late 1992 and early 1993. The decrease is especially noticeable in the mid-latitudes. They are unable to say, however, what the cause of this decrease is.

J. B. Kerr and C. T. McElroy of Environment Canada announce the first definitive data showing long-term increases in the level of ultraviolet radiation reaching

1993 *cont.* North America. Kerr and McElroy attribute this increase to depletion of the ozone layer.

Scientists at the National Oceanic and Atmospheric Agency (NOAA) report the lowest level of ozone over the Antarctic ever recorded. The level recorded is 88 dobson units on October 6. The width of the ozone hole is somewhat less than in previous years, but one vertical band of the stratosphere over the Antarctic between 13.5 kilometers and 19 kilometers from the Earth's surface is completely devoid of ozone.

A research team led by Dr. James W. Elkins of NOAA announces that the rate of increase of CFC-11 and CFC-12 in the atmosphere has begun to slow down. They predict that stratospheric concentration of these chemicals will level off before the year 2000.

E. I. du Pont de Nemours & Company announces that it will discontinue production of the chemicals by the end of 1994, a year ahead of its previously announced schedule.

Speaking at the 1993 International CFC and Halon Alternatives Conference, Paul Norris, president of Allied Signal Chemicals and Catalysts, announces that worldwide production of CFCs had fallen by more than 50 percent since 1986 and by 15 percent in the previous year alone. As an example of the way in which CFCs are being replaced by less harmful hydrofluorocarbons (HCFCs), he said that 90 percent of all new vehicle air-conditioners operated on HFC-134a, an alternative to CF_3CH_2F.

1994 An EPA rule banning the sale of "nonessential" products containing CFCs takes effect on 17 January.

A. R. Ravishankara and Carl Howard of the National Aeronautics and Space Administration announce that their research shows that hydrofluorocarbons have no harmful effects on the environment. The news is welcomed by representatives of industry as a possible alternative to CFCs that will soon be banned.

Scientists at NOAA announce research that suggests that fluorocarbons, possible replacements for CFCs, may

1994 *cont.* contribute to global warming. Not only do the fluorocarbons absorb radiation reflected from the Earth, but they may also remain in the Earth's atmosphere for up to 50,000 years.

The National Weather Service announces that it will begin publishing a daily forecast of ultraviolet light exposure. The forecast is necessary, the Service says, because of increased exposure to UV light caused by depletion of stratospheric ozone.

Dr. Max Bothwell, of the National Water Research Institute in Burlington, Ontario, reports that UVB radiation is causing unexpectedly severe damage to midge larvae. The report causes concern among scientists because midge larvae form an important component of the food chain for many freshwater fish. A number of authorities point to Bothwell's study as an additional reason for being concerned about depletion of stratospheric ozone.

Further studies of stratospheric ozone over the Antarctic are carried out during two research programs: the Airborne Southern Hemisphere Experiment (ASHE) and the Measurements for Assessing the Effects of Stratospheric Aircraft (MAESA) project.

NASA scientists report that its Total Ozone Mapping Spectrometer (TOMS) has apparently failed after fourteen-and-one-half years of reporting daily on stratospheric ozone concentrations.

Some of the best data ever obtained on the ozone layer is collected by astronauts aboard the *Atlantic* space shuttle flight of November 4–14. Some of these data provide convincing proof that human activities are responsible for the thinning of the ozone layer. That proof comes from the detection of the chemical hydrogen fluoride at a level that cannot be explained on the basis of natural sources alone. NASA Deputy Director Anne Douglas is quoted as saying "We have this thing nailed. There is no other possibility [other than human activities]."

1994 *cont.* Officials at the Alliance for Responsible Atmospheric Policy claim that millions of pounds of CFCs are being imported illegally into the United States each year. In addition to environmental problems caused by the CFCs, experts estimate that the importation of 20 million pounds of the chemicals each year are costing U.S. taxpayers about $100 million a year in lost excise taxes.

Measurements of the hole in the ozone layer over the Antarctic for 1994 indicate that the hole is about the same size as it was in 1993, and slightly less than its largest measured size of 9.4 million square miles, detected in 1992.

The first conviction obtained under the ozone protection provisions of the 1990 amendments to the Clean Air Act is announced. The owner of a car-repair shop in St. Louis admits to draining automotive air conditioners without proper training or equipment. As a result, the CFC coolant escaped into the air. The penalty for violating this law is a maximum of five years in prison and a $250,000 fine.

Biographical Sketches 3

Many individuals have participated in the discussion about depletion of the ozone layer, in both significant and modest ways. Brief biographical sketches of some of the most important of these individuals are provided in this chapter. Most of those listed here are from the field of science, although some made their contributions in the areas of politics, economics, or another field.

James G. Anderson (1944–)

James G. Anderson was born on 4 February 1944 in Spokane, Washington, but he grew up in nearby Pullman. He completed his B.S. degree in physics at the University of Washington in 1966 and then took his Ph.D. in physics and astrogeophysics at the University of Colorado in 1970.

Anderson says he was disappointed by the theoretical bent of his early physics courses, but he was delighted to be given the opportunity to build a spectrometer, a device for detecting the presence of elements in a material, during one of his summers at graduate school. After earning his doctoral degree, he continued this line of research as a postdoctoral fellow at the University of Pittsburgh from 1971 to 1972. There he constructed one of the first instruments for detecting the presence of free radicals in the stratosphere.

Anderson went on to become research assistant professor of physics at Pittsburgh from 1972 to 1975 and then research scientist in the Space Physics Research Laboratory at the University of Michigan from 1975 to 1978. After a brief term as associate professor in the Departments of Chemistry and Atmospheric and Oceanic Science at the University of Michigan, Anderson accepted an appointment as Robert P. Burden Professor of Atmospheric Chemistry at Harvard University. In 1982 he was appointed Philip S. Weld Professor of Atmospheric Chemistry at Harvard, a post he has held since.

It was Anderson's expertise in the area of sampling stratospheric air that in 1987 led to proof that CFCs are intimately involved in the destruction of ozone. Instruments designed by Anderson were carried aboard the ER-2 aircraft that flew through the stratosphere above the Antarctic and detected record-low concentrations of chlorine monoxide. Anderson's work is considered to have been a major impetus to decisions reached at the Montreal conference on ozone depletion later the same year. Two years later, Anderson provided some of the first evidence for the development of a hole in the ozone layer over the Arctic.

For his research, Anderson has received a number of honors and awards, including the American Chemical Society National Award for Creative Advances in Environmental Science and Technology (1989), the Ledley Prize for the Most Valuable Contribution to Science by a faculty member at Harvard, the Gustavus John Esselen Award for Chemistry in the Public Interest of the Northeastern Section of the American Chemical Society in 1993, and the E. O. Lawrence Award in Environmental Science and Technology in 1993. Anderson is the author or coauthor of more than 100 research papers.

Richard Elliot Benedick (1935–)

Richard Benedick was born in New York City on 10 May 1935. He received his B.A. *summa cum laude* from Columbia University in 1955, his M.A. in economics from Yale University in 1956, and his doctorate in business administration from Harvard University in 1962.

Benedick was hired as an economist by the U.S. Department of State in 1958 and held a number of positions with the department in the following four decades. He held assignments in Teheran, Karachi, Bonn, and Paris and was counselor for economic and commercial affairs at the American Embassy in Athens from 1975 to 1977. From 1979 to 1984, Benedick was coordinator of population affairs at the Department of State. In this position, he was responsible for carrying out U.S. policies on population, family planning, and biomedical research.

In 1984, Dr. Benedick was appointed deputy assistant secretary of state for environment, health and natural resources. In that position, he assumed responsibility for international aspects of U.S. activities and policies in a variety of fields, including climate change, AIDS, biotechnology, oceans, tropical forests, and wildlife conservation. In his State Department post, Benedick also became chief U.S. negotiator at the Vienna and Montreal meetings that resulted in international agreements to limit ozone-depleting substances. His book on the process by which these agreements were reached, *Ozone Diplomacy: New Directions in Safeguarding the Planet,* is one of the most interesting texts dealing with the politics of ozone depletion. As of 1994, the book had gone through its fourth printing at Harvard University Press.

In 1987, Benedick left his position with the State Department to become a senior fellow at the World Wildlife Fund. He also lectures, consults, and writes on environmental and population issues and serves on the board of a number of agencies and corporations. Since 1992, he has also been visiting professor at the Academie Internationale de l'Environment in Geneva. In addition to his book on the ozone issue, Benedick has written *The High Dam and the Transformation of the Nile, Industrial Finance in Iran,* and over 60 articles and chapters in books in U.S. and foreign publications.

John Chafee (1922–)

John Chafee was born in Providence, Rhode Island, on 22 October 1922. When the United States entered World War II, Chafee left Yale University to enlist in the Marine Corps. Upon his release from the service in 1946, Chafee returned to Yale, from which he received his B.A. in 1947. He then earned his LL.B. degree from Harvard Law School in 1950. During the Korean War, Chafee was recalled to serve in the Marine Corps. He was discharged as a captain in 1952.

Chafee entered politics in 1957, when he was elected as a Republican to the Rhode Island House of Representatives. He was reelected to two more terms and served as minority leader from 1959 to 1963. Chafee then ran for governor of the state and was elected by a slim 398-vote margin. He was then reelected in 1964 and 1966, both times by the largest margins in state history. In 1969 Chafee was asked to serve as secretary of the navy, a post he held for three and a half years.

In 1976 Chafee was elected to the U.S. Senate from Rhode Island, the first Republican to have achieved that distinction in 46 years. He was reelected in 1982, 1988, and 1994. During his long political career, Chafee has been particularly interested in problems of the environment. He has served for many years as ranking Republican on the Senate Committee on Environment and Public Works. In the late 1980s he was active in pushing for legislation that would ensure protection for the ozone layer. At one point, he introduced (with Senator Max Baucus) a bill that essentially reflected the position of the National Resources Defense Council on this matter.

Sydney Chapman (1888–1970)

Sydney Chapman was born in Eccles, Lancashire, England, on 29 January 1888. He earned his B.S. in engineering at Manchester University in 1907. He then moved to Cambridge University, from which he received a B.S. and M.S. in mathematics in 1908, a B.A. in mathematics in 1911, and a doctor of science degree in 1912.

Chapman's initial interests were in the area of kinetic theory of gases. He improved on James Clerk Maxwell's theory of thermal diffusion of gases and in the period 1912–1917 experimentally confirmed his own theory on that subject. After that time, Chapman's interests turned to the atmosphere. He studied a variety of phenomena, including auroras, atmospheric tides, and magnetic properties of the atmosphere.

The work for which Chapman is best known today was published in 1931. In two now-famous papers, he hypothesized that solar radiation causes the conversion of oxygen atoms in the stratosphere into ozone molecules. He then showed that a steady state exists between the creation and destruction of ozone molecules and that the series of chemical reactions involved in this steady state accounts for the unexpected warmth of the upper stratosphere. In his honor, the ozone layer is sometimes referred to as the *Chapman layer,* and the series of reactions by which that layer is maintained is known as the *Chapman reactions.*

Throughout his long and illustrious professional career, Chapman held a number of prestigious appointments. He served as chief assistant at the Royal Observatory in Greenwich (1910–1914 and 1916–1918), lecturer at Trinity College, Cambridge

(1914–1919), professor of mathematics and natural philosophy at Manchester University (1919–1924), chief professor of mathematics at the Imperial College, London (1924–1946), Sedleian Professor of Natural Philosophy (1946–1953) and later professor emeritus at Oxford University, member of the research staff at the High Altitude Observatory in Boulder, Colorado (1955–1970), visiting professor and advisor to the scientific director at the Geophysical Institute of Alaska, senior research scientist at the Institute of Science and Technology and the University of Michigan, and special visiting professor at the University of Colorado (1965).

Chapman's list of honors and awards is extensive. It includes the Adams Prize (1929), the Feltrinelli International Prize of the National Academii dei Lincei of Rome (1956), the Bowie Medal of the American Geophysical Union (1962), and the Hodgkins Medal and Prize of the Smithsonian Institution (1965). He was elected to the Symons Memorial lectureship of the Royal Meteorological Society (1926), the Rouse Ball lectureship at Cambridge (1929), the Bakerian lectureship of the Royal Society (1931), and the Halley lectureship at Oxford (1943).

Chapman died at Boulder, Colorado, on 16 June 1970.

Ralph Cicerone (1943–)

Ralph Cicerone was born in New Castle, Pennsylvania, on 2 May 1943. He received his undergraduate education at the Massachusetts Institute of Technology, from which he received his B.S. in 1965. He then went on to earn an M.S. and a Ph.D. (in electrical engineering and physics) at the University of Illinois in 1967 and 1970, respectively.

After earning his doctorate, Cicerone accepted an appointment as an associate research scientist in aeronomy at the University of Michigan's Space Physics Research Laboratory. One of his first research projects at Michigan was a contract awarded to him and fellow faculty member Richard Stolarski by the National Aeronautics and Space Administration (NASA). NASA was interested in finding out about the atmospheric effects that might result from the launch of the agency's new space shuttles. As a result of their research, Cicerone and Stolarski discovered that chlorine and chlorine compounds might have profound effects on ozone found in the stratosphere. Their 1974 paper on this subject was one of the first to raise the question of possible effects on the ozone layer resulting from human activities.

Cicerone remained at Michigan until 1978 when he accepted an appointment at the Scripps Institute of Oceanography of the

University of California at San Diego. In 1980 he left Scripps to become senior scientist and director of the Atmospheric Chemistry Division of the National Center for Atmospheric Research at Boulder, Colorado. In 1989 he returned to California to take the post of Daniel G. Aldrich Professor and chair of the Geosciences Department at the University of California at Irvine.

Paul Crutzen (1933–)

Paul Crutzen was born in Amsterdam on 3 December 1933. In 1954 he received his degree in civil engineering and then took a position with the Bridge Construction Bureau of the City of Amsterdam. From 1956 to 1958, Crutzen served with the Dutch military service.

At the conclusion of his military service, Crutzen moved to Sweden and took a job with the House Construction Bureau in Gävle. At the same time, he enrolled at the University of Stockholm to begin his graduate studies. He later earned master of science (1963) and doctor of philosophy (1968) degrees at Stockholm. During this period, Crutzen was also working in various research capacities in the Department of Meteorology at the university.

From 1969 to 1971 Crutzen was a postdoctoral fellow of the European Space Research Organization at the Clarendon Laboratory of the University of Oxford. During this time, he was also working on his doctor of science degree, which was awarded with highest possible distinction in 1973. Crutzen then moved to the United States, where he was employed at the National Center for Atmospheric Research (NCAR) and the Environmental Research Laboratories of the National Oceanic and Atmospheric Administration in Boulder, Colorado. He became director of the Air Quality Division at NCAR in 1977. From 1976 to 1981 Crutzen was also adjunct professor of atmospheric sciences at Colorado State University.

In 1980, Crutzen returned to Europe to take the post of director of the Atmospheric Chemistry Division of the Max-Planck-Institute for Chemistry in Mainz, Germany. Three years later he was promoted to executive director of the institute. While at Mainz, Crutzen has also served as professor at the Universities of Chicago (1987–1991) and California, La Jolla (1992–).

Crutzen's special field of research interest has long been atmospheric chemistry and its role in biological and geological cycles and in climate change. As early as 1972 he was raising questions about the possible effects of supersonic aircraft on the composition of the stratosphere. His work on the effects of so-called solar proton events on the formation and destruction of nitrogen compounds in the atmosphere has been especially important in the development of theories about ozone destruction.

In addition to nearly 150 refereed papers, he has coauthored three books and edited three more books. Among his many awards are the Rolex-Discover Scientists of the Year (1984), the Leo Szilard Award for Physics in the Public Interest of the American Physical Society (1985), the Tyler Prize for the Environment(1989), the Volvo Environmental Prize (1991), and the German Environmental Prize of the Federal Foundation for the Environment (1994).

Gordon Miller Bourne Dobson (1889–1976)

Gordon Dobson was born at Knott End, Windermere, England, on 25 February 1889. Dobson attended Sedbergh School and then enrolled at Gonville and Caius College at Cambridge University. He earned first class honors in the natural sciences at Gonville and Caius in 1910.

Dobson showed an early interest in and aptitude for the sciences and had his first paper published in the journal *Nature* in 1911. That paper came to the attention of Sir W. Napier Shaw, director of the Meteorological Office, who then offered Dobson a position at the Kew Observatory in London. Two years later, Dobson accepted an appointment as meteorological advisor at the newly established Military Flying School on Salisbury Plain. There he began pioneering work with balloons for the measurement of wind and atmospheric conditions. During World War I, Dobson was director of experimental research at the Royal Aircraft Establishment. At the war's conclusion, he was appointed university lecturer in meteorology at Oxford University.

It was at Oxford that Dobson made his first major discoveries about the atmosphere. In 1922 he published a paper showing the presence of a warm layer in the atmosphere at an altitude of about 50 kilometers. This report was startling since scientists had long believed that the temperature of the atmosphere decreases continuously from the Earth's surface upward. In recognition of this work, Dobson was awarded a doctor of science degree from Oxford in 1924.

Dobson hypothesized that the warm layer he had discovered in the stratosphere was probably caused by the absorption of ultraviolet radiation by ozone molecules. He devoted the greater part of the rest of his professional career to studying the ozone layer in more detail. Over the next 30 years he developed a series of increasingly more sophisticated instruments with which to measure the concentration of ozone in the stratosphere. In recognition of this work he received a number of awards and honors, including election to a fellowship at Merton College, to a fellowship in the Royal Society, and to a readership in meteorology at Oxford, all in 1927. He was also awarded the Rumford Medal of the Royal Society in 1942 and was named full professor at Oxford in 1945. In his honor, the unit of measurement of stratospheric ozone concentration has been named the *dobson.*

Dobson died on 10 March 1976.

Anne Gorsuch [Burford] (1942–)

Anne Gorsuch Burford was born Anne McGill in Casper, Wyoming, on 21 April 1942. She received her B.A. and LL.B. degrees from the University of Colorado in 1961 and 1964 respectively.

Over the next two decades, Gorsuch held a number of positions, including trust administrator with the First National Bank of Denver, instructor at Metropolitan State College in Denver, assistant district attorney in Jefferson County, Colorado, and deputy district attorney in Denver. Gorsuch was elected a member of the Colorado House of Representatives in 1977, eventually serving two terms. While in the Colorado House, she was a leader in the effort to reduce government legislation and regulation in the areas of environmental issues.

In 1981, President Ronald Reagan selected Gorsuch to be administrator of the Environmental Protection Agency. During her two-year tenure at the EPA, Gorsuch promoted a go-slow policy on ozone-related issues. She argued that clear evidence for the loss of ozone and the role of CFCs in the problem was not yet available. To the extent that an ozone problem did exist, she argued that free-market policies would be the best way of dealing with such problems.

Just prior to resigning as EPA administrator, Gorsuch married Robert Burford, Reagan's director of the Bureau of Land Management.

David J. Hofmann (1937–)

David Hofmann was born in Albany, Minnesota, on 3 January 1937. After graduating from high school in 1954, he enlisted in the U.S. Navy. He left the Navy in 1958 and entered the University of

Minnesota, from which he eventually received his B.S. in physics (with distinction) in 1961, his M.S. degree in 1963, and his Ph.D. in physics in 1966.

Hofmann was associated with the University of Wyoming for more than two decades. He was assistant professor of physics (1966–1970), associate professor of physics (1970–1975), and finally full professor of physics (1975–1991) at Wyoming. He was also head of the Department of Physics and Astronomy from 1978 to 1983. In 1990 he accepted an appointment as chief scientist of the Climate Monitoring and Diagnostics Laboratory at the National Oceanic and Atmospheric Administration's Environmental Research Laboratory in Boulder, Colorado, a position he continues to hold today. Since May 1991 he has also served as adjoint professor in the Department of Astrophysical, Planetary and Atmospheric Sciences of the University of Colorado in Boulder.

Professor Hofmann has been involved in balloon research of the troposphere and stratosphere for more than 30 years. His specialty within that field has been stratospheric aerosols, tiny droplets of liquids and solids found in the lower stratosphere. Hofmann's research has provided some key pieces of information about the question of ozone depletion. In 1976, for example, he was able to provide evidence that chlorofluorocarbons are broken apart in the atmosphere by solar radiation. His work during the National Ozone Expedition provided one of the first and most detailed vertical profiles of ozone concentration above the Antarctic, showing that at some altitudes that concentration dropped nearly to zero.

Among his many honors and awards are the University of Wyoming College of Arts and Sciences Distinguished Faculty Award (1975), the United States Antarctic Service Medal (1979), the Humboldt Prize from the Federal Republic of Germany (1982), and the U.S. Department of Commerce Distinguished Authorship Award (1990). He has more than 150 published papers to his credit.

Harold Sledge Johnston (1920–)

Harold Johnston was born in Woodstock, Georgia, on 11 October 1920. After earning his B.A. from Emory University in Atlanta in 1941, Johnston continued his education at the California Institute of Technology (Cal Tech), from which he received his Ph.D. in 1948. He served as instructor, assistant professor, and associate professor at Stanford University from 1947 to 1956 and then as associate professor of chemistry at Cal Tech for one more year, from 1956 to 1957. Johnston was then appointed professor of chemistry at the University of California at Berkeley. He later served as dean of the College of Chemistry from 1966 to 1970.

While still a graduate student at Cal Tech, Johnston had worked on the chemical reactions involved in the production of photochemical smog in areas such as Los Angeles. He found that complex interactions between oxides of nitrogen and ozone were responsible for some of the worst features of this type of smog. Two decades later he was drawn back into this line of research when plans for the construction of supersonic (SST) aircraft were announced. Johnston calculated that a fleet of SSTs could release enough oxides of nitrogen into the atmosphere to reduce ozone levels by as much as 23 percent.

By the mid-1970s Johnston had become one of the world's authorities on ozone-depletion problems. When F. Sherwood Rowland and Mario Molina announced their theory of ozone depletion in 1974, Johnston was one of the first atmospheric scientists to independently confirm their predictions.

James Bews Kerr (1946–)

James Kerr was born in Toronto on 7 December 1946. He received his bachelor of science degree with a major in physics from the University of British Columbia in 1968. He then returned to the University of Toronto, where he was awarded his master of science (1970) and doctor of philosophy (1973) degrees, both in atmospheric physics. Upon receiving his Ph.D. degree, Kerr accepted a position with the Atmospheric Environment Service of the Canadian government, a position he has held ever since. He has held the titles of research meteorologist (1973–1975), research scientist (1975–present), and head of Ozone Research and Monitoring (1990–present).

Kerr has made some important contributions to the development of instruments for the detection and measurement of ozone and UVB radiation. Best known of these is the Brewer ozone spectrophotometer, of which more than 100 are now in use throughout the world. He is also the coinventor of the UV Index, now used in Canada, the United States, and a number of other nations.

Kerr is the author of more than 90 research papers and articles dealing with the ozone layer and ultraviolet radiation. Among the most important of these was his November 1993 paper with C. T. McElroy documenting a significant increase in UVB levels above Toronto over the preceding decade.

Margaret Kripke (1943–)

Margaret Kripke was born Margaret Louise Cook in Concord, California, on 21 July 1943. She attended the University of California at Berkeley, where she was awarded her A.B. (1965) and M.A. (1967) in bacteriology and her Ph.D. (1970) in immunology.

Over the next decade, Kripke held an increasingly important series of appointments as research associate at the Ohio State University (1970–1972) and at the University of Louisville School of Medicine, assistant professor of pathology at the University of Utah College of Medicine, senior principal scientist and then director of the Immunology and Cancer Biology Program at the National Cancer Institute's Frederick Cancer Research Facility in Frederick, Maryland. Finally, in 1979 she was appointed Vivian L. Smith Professor and chair of the Department of Immunology of the M. D. Anderson Cancer Center of the University of Texas in Houston.

Kripke has taken a leadership role among biological and medical scientists in warning about possible health effects of UVB radiation resulting from destruction of the ozone layer. As an authority on skin cancer, she has discussed the potential threat of ozone depletion for this disease, but she has also argued that increased levels of UVB radiation can have effects in other areas of human health.

In 1993, Kripke was chosen president of the American Association for Cancer Research.

Arlin Krueger (1933–)

Arlin Krueger was born in Lamberton, Minnesota, on 22 October 1933. He was educated at the University of Minnesota, from which he received his bachelor of arts degree in physics in 1955, and the Colorado State University, from which he received his Ph.D. degree in atmospheric sciences in 1984. For the first decade of his academic career, Krueger worked at the Naval Weapons Center in China Lake, California. There he worked on the development of ozone detection systems using balloons and rockets. His work there resulted in the development of the first devices for measuring ozone concentrations with artificial satellites.

In 1969, Krueger accepted an appointment with the National Aeronautic and Space Administration's (NASA) Goddard Space Flight Center, where he is a senior research scientist in the Atmospheric Chemistry and Dynamics Branch of the Laboratory for Atmospheres, Earth Sciences Directorate.

In 1972, Krueger suggested the development of the Total Ozone Mapping Spectrometer (TOMS) that was to become one of the most powerful tools in the detection and measurement of stratospheric ozone. The first TOMS was placed on the *Nimbus 7* satellite that was launched in October 1978. One of the major discoveries made with the TOMS device was the presence of sulfur dioxide emitted during the El Chichón volcanic eruption in 1982. Krueger's more recent work has been devoted to the development of TOMS instruments for forthcoming Earth Probe and ADEOS satellites. Krueger was awarded NASA's Exceptional Scientific Achievement Medal in 1991.

James Ephraim Lovelock (1919–)

James Lovelock was born on 26 July 1919 in Letchworth Garden City, England. He attended the University of Manchester, from which he received his bachelor of science degree in chemistry in 1941. He then earned a Ph.D. degree in medicine from the London School of Hygiene and Tropical Medicine (1948) and a D.Sc. degree in biophysics from London University (1959).

Lovelock's first job was with the Medical Research Council at the National Institute for Medical Research in London. Then, in 1946, he took a position at the Common Cold Research Unit at Harvard Hospital in Salisbury, Wiltshire. Lovelock has held two Rockefeller fellowships, the first spent at Harvard University from 1954 to 1955 and the second at Yale University from 1958 to 1959. He 1961 he resigned from the National Institute in London to accept an appointment as professor of chemistry at Baylor University College of Medicine in Houston. During his three-year appointment in Texas, he also worked at the Jet Propulsion Laboratory in Pasadena, California, on lunar and planetary research.

In 1964, Lovelock resigned his position at Baylor and began a career as an independent scientist. He is perhaps best known today for the Gaia hypothesis, a concept suggesting that the Earth is a self-regulating system that has developed the ability to maintain a climate and chemical composition that is optimal for living organisms. In 1957, Lovelock invented the electron capture detector, a device used to measure the abundance of chlorofluorocarbons and nitrous oxide in the atmosphere.

When Rowland and Molina first proposed their hypothesis about the role of CFCs in ozone depletion, Lovelock expressed serious doubts about the concept. He felt that the early popular reaction to the Rowland-Molina theory, in particular, was more pronounced than it need have been. Since the end of the 1980s, however, he has come to the conclusion that CFCs should, indeed, be banned from production and use.

Jerry D. Mahlman (1940–)

Jerry Mahlman was born in Crawford, Nebraska, on 21 February 1940. He attended Chadron State College, where he earned his bachelor of arts degree in 1962, and Colorado State University, from which he received his master of science degree in 1964 and his Ph.D. in 1967. After graduating from Colorado State, Mahlman was appointed assistant professor of meteorology at the U.S. Naval Postgraduate School in Monterey, California. He then went on to take a position with the Geophysical Fluid Dynamics Laboratory (GFDL) of the National Oceanic and Atmospheric Administration in Princeton, New Jersey. At GFDL Mahlman has held the posts of research meteorologist (1970–1978), senior research meteorologist (1978–1984), and, finally, director of the laboratory (1984–present).

Mahlman has long been interested in efforts to understand the behavior of the troposphere and stratosphere. Much of his time has been spent on the development of mathematical models that explain the interactions among chemical, radiative, dynamical, and transport factors in the atmosphere. Although his primary interest has been on climate change, he has also carried out some fundamental research on stratospheric ozone depletion. In the late 1980s, he developed a hypothesis that explained ozone depletion at least partially as the result of changes in polar stratospheric clouds.

Mahlman has served on or chaired more than 30 important committees and panels. He is also the author or coauthor of more than 80 publications. Among the awards and honors he has received are the Distinguished Service Award of Chadron State College (1984), the Gold Medal of the Department of Commerce (1986), selection as the First Annual Jule G. Charney Lecturer (1993), the Carl-Gustav Rossby Research Medal of the American Meteorological Society (1994), and the Presidential Distinguished Rank Award (1994).

Michael McElroy (1939–)

Michael McElroy was born in Northern Ireland on 18 May 1939. He was educated at Queen's University in Belfast, Northern Ireland, from which he received his B.A. (with honors) in applied mathematics in 1960, and his Ph.D. in applied mathematics in 1962. After graduation, McElroy spent a year at the Theoretical Chemistry Institute of the University of Wisconsin (1962–1963).

He then accepted a job in the Planetary Sciences Division at the Kitt Peak National Observatory in Tucson, Arizona. He was assistant physicist (1963–1965), associate physicist (1965–1967), and physicist (1967–1970) during his tenure there.

In 1970, McElroy was appointed Abbott Lawrence Rotch Professor of Atmospheric Sciences in the Division of Applied Sciences at Harvard University. He still holds that post, and since 1986 he has also been chair of the Department of Earth and Planetary Sciences at Harvard.

Along with his colleague, Steven Wofsy, McElroy published one of the seminal papers in the history of ozone-depletion research. In 1975, McElroy and Wofsy published a paper confirming the Rowland-Molina theory relating ozone destruction to the presence of chlorine in the stratosphere. At first, the Rowland-Molina, McElroy-Wofsy discovery seemed to be relatively unimportant since it arose out of studies of possible flight in the stratosphere. As the role of CFCs became more obvious, however, that discovery took on greater importance.

McElroy has served on a number of major commissions, committees, and study groups investigating the problem of ozone depletion. Included among these are the Stratospheric Research Advisory Committee and the Space and Earth Science Advisory Committee of the National Aeronautics and Space Administration, the Space Program Advisory Panel of the Office of Technology Assessment, and the Committee on the Atmospheric Effects of Nuclear Explosions and the Committee for the International Geosphere/Biosphere Program of the National Academy of Sciences.

Thomas Midgley, Jr. (1889–1944)

Thomas Midgley, Jr., was born in Beaver Falls, Pennsylvania, in 1889. He received a bachelor's degree in engineering from Sibley College of Engineering at Cornell University.

Midgley's first job was with the National Cash Register Company in Dayton, Ohio. After five years with NCR, however, he accepted an offer to work with the famous Charles Kettering in a small invention firm also located in Dayton. At Kettering's firm in December 1921, Midgley made the first of two inventions for which he is famous, the gasoline additive tetraethyl lead. His discovery of this commercially critical antiknock compound won him fame and a modest fortune.

Seven years later, Midgley made a second famous discovery: chlorofluorocarbons. The discovery came as the result of efforts to find a safe, nontoxic, odorless, colorless gas that could be used as a refrigerant. Midgley's discovery was patented on 31 December 1928, although an official announcement of his work was not made for another year.

Midgley's two inventions brought him honors and awards from academic institutions, professional organizations, and government agencies. For example, he received the American Chemical Society's prestigious Priestly Prize in 1941.

At the age of 51, Midgley was struck down with an attack of poliomyelitis that left him confined to a wheelchair. Although he maintained his usual pace of business for a few years, he eventually became despondent about his physical condition and eventually committed suicide on 2 November 1944.

Alan S. Miller (1949–)

Alan Miller was born in Birmingham, Michigan, on 22 December 1949. He attended Seaholm High School in Birmingham, where he was a member of the debate and track teams. Miller then enrolled at Cornell University, from which he received a bachelor's degree in government in 1971. His interests in environmental and peace issues date to his college years at Cornell. After leaving Cornell, Miller continued his education at the University of Michigan, where he earned both a master's degree in public policy and a J.D. from the Law School in 1974.

Miller's first work experience after college was with the Environmental Law Institute in Washington, D.C., where he was employed from 1974 to 1977. He was then awarded a Fulbright Scholarship to study at the Macquarie University in North Ryde, Australia. Upon his return to the United States, Miller served briefly as visiting assistant professor at the University of Iowa College of Law in Iowa City.

Since mid-1979, Miller has held a variety of posts with environmental organizations, including staff attorney at the Natural Resources Defense Council (NRDC; 1978 to 1984), associate at the World Resources Institute in Washington, D.C. (1984 to 1986) and, most recently, executive director of the Center for Global Change at the University of Maryland. Simultaneously with these jobs, Miller has also served as visiting professor at a number of institutions, including the Washington College of Law of American University (Fall 1986), Widener University Law School in Wilmington, Delaware (1988–1989), Vermont Law School (summers of 1991, 1992, and 1993), Duke Law School Summer Faculty in Brussels (1993), and Maryland Law School (1989 to present).

During his years at NRDC, Miller was especially active in efforts to protect the ozone layer. In 1980, for example, he wrote a detailed commentary on the status of scientific information about the effects of CFCs on ozone depletion. Over the next four years, he testified before Congress, prepared reports, and worked actively in other ways to bring about a ban on the production and use of CFCs as an essential step in protecting the ozone layer from further deterioration. Miller is the author or coauthor of more than two dozen books and articles, including *Green Gold: The Race to Capture Industrial Dominance in the 21st Century* (with C. Moore), *Environmental Regulation: Law, Science and Policy* (with R. Percival, C. Schroeder, and J. Leape), and *The Sky is the Limit: Strategies To Protect the Ozone Layer* (with I. Mintzer).

Mario J. Molina (1943–)

Mario Molina was born in Mexico City on 19 March 1943. He graduated from the Academia Hispano Mexicana in Mexico City in 1959 and then earned his degree in chemical engineering from the Universidad Nacional Autonoma de Mexico in 1965. Molina then spent a year at the University of Freiburg before enrolling at the University of California at Berkeley for his doctoral studies. He was awarded his Ph.D. in physical chemistry from Berkeley in 1972.

Molina was employed as assistant professor at the Universidad Nacional Autonoma for one year between his year at Freiburg and his doctoral studies at Berkeley. After receiving his Ph.D., he become a research associate at Berkeley for one year before accepting a similar position at the University of California at Irvine. He remained at Irvine from 1973 to 1982, working his way up to associate professor. From 1983 to 1989 he served as senior research scientist at the Jet Propulsion Laboratory in Pasadena, California. In 1989, Molina accepted a joint appointment as professor in the Departments of Chemistry and of Earth, Atmospheric, and Planetary Sciences at the Massachusetts Institute of Technology (MIT) in Cambridge, Massachusetts. He now holds the title of Martin Professor of Atmospheric Chemistry at MIT.

Molina's name will always be associated with a historic paper that he wrote in 1974 with F. Sherwood Rowland regarding the effects of chlorine on ozone in the stratosphere. The paper set off a discussion of and debate over the causes of ozone depletion that continues to the present day. Molina has continued to work on various aspects of the ozone depletion problem ever since.

Among the many awards that Molina has received are the Tyler Award in 1983, the Esselen Award of the American Chemical Society in 1987, and the Newcomb-Cleveland Prize of the American Association for the Advancement of Science.

Auguste Piccard (1884–1962)

Auguste Piccard was born in Basel, Switzerland, on 28 January 1884. His father, Jules Piccard, was chairman of the Department of Chemistry at the University of Basel. After graduation from the local high school, Piccard entered the Federal Institute of Technology in Zurich where he earned his bachelor of science and doctoral degrees in mechanical engineering. Between 1907 and 1920, Piccard taught in Zurich. He then accepted an appointment as professor of physics at the Brussels Polytechnic Institute, a post he held until his retirement in 1954.

Piccard earned his fame by exploring higher into the Earth's atmosphere and deeper into its oceans than had any person before him. Prior to the 1930s, almost all atmospheric research had been conducted with relatively unsophisticated instruments on unmanned balloons. Piccard's view was that unmanned ascents could never provide the quality of data that could be obtained from balloons in which humans could travel. He resolved, therefore, to design a pressurized gondola in which observers could travel well beyond the 29,000 foot level that had marked the previous barrier to manned flight. In 1931, Piccard and a colleague traveled in a balloon to an altitude of about ten miles, more than three miles higher than any human had ever gone before. As a result of this achievement, scientists obtained some of the most important firsthand data about the stratosphere that had yet come to them.

Later in life Piccard became interested in oceanographic research. In the 1930s he constructed one of the most effective devices for exploring the ocean depths: the bathyscaphe. Piccard died in Lausanne, Switzerland, on 24 March 1962.

Veerabhadran Ramanathan (1944–)

Veerabhadran Ramanathan was born in Madras, India, on 24 November 1944. He earned his bachelor's degree from Annamalai University in 1965, his master's at the Indian Institute of Science in Bangalore in 1970, and his doctorate in atmospheric sciences from the State University of New York at Stony Brook in 1974.

After graduating from Stony Brook, Ramanathan spent a year as a research fellow at NASA's Langley Research Center. He then took a position with the National Center for Atmospheric Research (NCAR), in Boulder, Colorado, an affiliation he has maintained for more than two decades. At NCAR Ramanathan has been leader of the Cloud Climate Interactions Group since 1981 and senior scientist since 1982. He has also held faculty appointments at Colorado State University at Fort Collins since 1985 and the University of Chicago since 1986.

While still in India, Ramanathan had developed an interest in chlorofluorocarbons. One of his tasks as an industrial refrigeration engineer there was to find ways to prevent CFCs from leaking from cooling systems.

While at Stony Brook, Ramanathan turned once again to the problem of CFCs. There he became particularly interested in the ability of CFCs to absorb infrared radiation reflected from the Earth's surface, in much the same way that carbon dioxide does. In September 1975, he published a paper pointing out that CFCs were, in addition to ozone depleters, greenhouse gases also. That is, their presence in the atmosphere increases the amount of infrared radiation reflected from the Earth that is captured by the atmosphere and, therefore, helps raise the atmosphere's annual average temperature. Ramanathan's discovery meant that CFCs were an environmental problem not only because of their tendency to deplete stratospheric ozone but also because of their potential for contributing to global warming.

F. Sherwood Rowland (1927–)

Sherwood Rowland was born in Delaware, Ohio, on 28 June 1927. He received his A.B. degree from Ohio Wesleyan University in 1948, his M.S. from the University of Chicago in 1951, and his Ph.D. from Chicago a year later. His first academic appointment was as instructor of chemistry at Princeton University from 1952 through 1956. He then took a position as assistant professor of chemistry at the University of Kansas, eventually working his way up to associate professor in 1958 and full professor in 1963.

In 1964, Rowland accepted an appointment as professor of chemistry at the University of California at Irvine. He became Aldrich Professor of Chemistry in 1985 and then Bren Professor of Chemistry in 1989. He has also served as Humboldt Senior Scientist in the Federal Republic of Germany (1981) and visiting scientist at the Japanese Society for the Promotion of Science (1980).

In 1974, Rowland published one of the seminal papers regarding ozone depletion with one of his assistants, Mario Molina. In that paper, Rowland and Molina argued that the build-up of CFCs in the stratosphere was likely to result in serious damage to the ozone layer. For more than two decades, that paper has served as a building block for theories about ozone destruction as well as a catalyst for debate over the seriousness of the problem.

Among Rowland's many honors have been the Tyler World Prize in Environmental Achievement (1983), the Silver Medal of the Royal Institute of Chemistry (1989), the Japan Prize in Environmental Science (1989), and the Dickson Prize of Carnegie-Mellon University.

William D. Ruckelshaus (1932–)

William D. Ruckelshaus was born in Indianapolis, Indiana, on 24 July 1932. He joined the U.S. Army as a private in 1953 and was released two years later with the rank of sergeant. He then enrolled at Princeton University, from which he received his bachelor of arts degree, *cum laude,* in 1957. Following his graduation from Princeton, Ruckelshaus entered Harvard Law School and earned his LL.B. there in 1960.

Ruckelshaus began a long and distinguished career in government in 1960 when he accepted an appointment as Deputy Attorney General for the state of Indiana. He also served as chief counsel in the office of the State Attorney General from 1963 to 1965. In 1967, Ruckelshaus ran for the Indiana House of Representatives, was elected, and was chosen majority leader.

Two years later, he was asked by President Richard Nixon to serve as assistant attorney general and head of the Civil Division of the Justice Department. In 1970, he was appointed by Nixon to head the newly created Environmental Protection Agency. Although he was part of an administration that was not particularly notable for its environmental record, Ruckelshaus was unusually sensitive to growing concerns over the possible effects of human activities on the ozone layer. He cooperated with environmental groups to encourage further research on this topic and began to implement government decisions that would reduce possible damage to the ozone layer. His record in this area stands in stark contrast with the more obstructionist approach taken by his successors in the EPA.

In 1973, Ruckelshaus left the EPA to become acting director of the FBI briefly and then deputy attorney general. He served in that post only 24 days. When he (along with his superior, Attorney General Elliot Richardson) refused to fire Watergate special prosecutor Archibald Cox in the infamous "Saturday Night Massacre" of 20 October 1973, Ruckelshaus resigned his post at Justice.

After leaving Washington, Ruckelshaus became senior vice president for law and corporate affairs at the Weyerhaeuser Corporation in Tacoma, Washington. Since 1988 he has been chairman of the board and chief executive officer at Browning-Ferris Industries, Inc., in Houston.

Christian Schönbein (1799–1868)

Christian Schönbein was born in Metzingen, Würzburg, on 18 October 1799. He was largely self-taught since his family was too poor to allow him to attend formal schools. As a young man, he was employed at a pharmaceutical factory in his home town and attended classes on an irregular basis at the Universities of Erlangen and Tübingen. Although he never earned a formal college degree, he was awarded an honorary doctorate by the University of Basel in 1828. He was then offered a faculty position there and eventually became full professor in 1835.

Schönbein has two great accomplishments to his credit. The first was his discovery in 1840 of ozone. He first identified the gas because of its sickeningly sweet smell. In fact, he suggested the name for the new discovery by using the Greek word, *ozein,* meaning "to smell."

The second great discovery for which Schönbein is known is the discovery of guncotton. The story is told that he found out about this powerful explosive when he attempted to clean up some spilled nitric acid with a piece of cotton cloth. The compound that was formed when the acid reacted with the cotton proved to be a powerful explosive. It eventually became a great improvement over the previously most popular explosive, gunpowder.

Schönbein died in Sauersberg, Baden, on 29 August 1868.

Susan Solomon (1956–)

Susan Solomon was born in Chicago on 19 January 1956. She received her bachelor of science degree in chemistry with high honors from the Illinois Institute of Technology in 1977 and then attended the University of California at Berkeley. There she was granted her master of science (1979) and Ph.D. in chemistry

(1981). Since receiving her doctoral degree, Solomon has been employed as a research scientist at the Aeronomy Laboratory of the National Oceanic and Atmospheric Administration in Boulder, Colorado.

Solomon is widely regarded as one of the world's authorities on the chemistry of atmospheric gases. She was chosen to head the National Ozone Expedition at McMurdo Station in Antarctica in 1986 and 1987, when she was only 30 years old. She has also served on a number of prestigious government committees that are typically not open to individuals as young as she.

Among the honors bestowed on Solomon have been the J. B. MacElwane Award of the American Geophysical Union, the Department of Commerce Gold Medal for Exceptional Service, and the Henry G. Houghton Award of the American Meteorological Society. In 1992, R&D magazine selected Solomon as scientist of the year. She currently is head of the Middle Atmosphere group of NOAA's Aeronomy Laboratory.

L. P. Teisserenc de Bort (1855–1913)

L. P. Teisserenc de Bort was born in Paris on 5 November 1855. He was frail and sickly as a child and did not complete a traditional course of education. He was, however, creative and imaginative in designing equipment and experiments, and at the age of 18 designed a new observatory to be built at Grasse. Seven years later he accepted an appointment at the Central Meteorological Bureau in Paris and, two years later, was promoted to chief meteorologist.

Four years later, unhappy with the limited research budget provided the Bureau, Teisserenc de Bort resigned his post and decided to use his own money to continue research on the atmosphere. For the rest of his life, he concentrated on developing techniques for the study of the atmosphere using unmanned balloons. His most important discoveries were summarized in a report he presented to the French Academy of Sciences in 1902. Among those discoveries was the first recognition that temperatures in the atmosphere decrease at a regular rate up to an altitude of about 11 kilometers. Beyond that point, and to the highest altitude Teisserenc de Bort could go, atmospheric temperatures seemed to remain constant.

This discovery prompted Teisserenc de Bort to conclude that the atmosphere consists of two layers. The lower of these he named the troposphere, meaning "sphere of change." At the upper limit of the troposphere, he said, was a break in temperature changes, a region for which he suggested the name tropopause, or "end of change." He then hypothesized that the region above the tropopause was characterized by layers of gases arranged according to their densities. He suggested the name stratosphere—"sphere of layers"—for this upper region.

Teisserenc de Bort died in Cannes on 2 January 1913.

Lee Thomas (1944–)

Thomas was born in Ridgeway, South Carolina, on 13 June 1944. He earned his bachelor of arts degree from the University of the South in 1967 and his master's in education from the University of South Carolina in 1971.

Thomas has spent most of his adult life in government and public service. From 1972 to 1977 he was executive director of the Criminal Justice Program in the governor's office in South Carolina. He then worked as an independent consultant in the field of criminal justice planning for two years before returning to government service as director of public safety programs for the state of South Carolina. In 1981, Thomas moved to Washington, D.C., to become associate director for state and local programs and support of the Federal Emergency Management Agency (FEMA). A year later he was appointed deputy director of the agency.

In 1983, President Ronald Reagan appointed Thomas assistant administrator of the Environmental Protection Agency (EPA), a post he held for two years. Then, in 1985, he was selected to replace Ann Gorsuch Burford as administrator of the EPA.

The direction taken by Thomas in dealing with ozone issues was quite different from that of his predecessor and probably quite different from what most observers had expected. He quickly came to accept the fact that ozone depletion was a real and serious problem that demanded the nation's attention. He met with representatives of various environmental groups, especially the National Resources Defense Council, in order to work out a national policy for dealing with the ozone issue. Overall, Thomas eventually received high marks from those concerned about ozone depletion for his manner of dealing with the problem.

Upon leaving government service in 1989, Thomas became affiliated with Law Environmental Inc. in Atlanta, Georgia. He is now chairman and chief executive officer of the organization.

Mostafa Tolba (1922–)

Tolba was born in Zifta, Egypt, in the delta of the Nile River, on 8 December 1922. He attended Cairo University, from which he earned a Special Bachelor of Science degree in botany with first class honors in 1943. He then continued his studies at the Imperial College of the University of London, receiving his Ph.D. degree in plant pathology there in 1949. For the ten-year period between 1949 and 1959, Tolba served as lecturer and assistant professor of botany at Cairo University and as professor of botany at Baghdad University.

In 1959, Tolba was asked to become assistant secretary-general of the National Science Council of Egypt, and two years later he was promoted to secretary-general of the organization. He went on to become cultural counselor and director of the Egyptian Education Bureau at the Egyptian Embassy in Washington (1963–1965), professor of microbiology at the National Research Center in Cairo, and the Egyptian under secretary of state for higher education (1965) and the minister of youth in 1971.

Tolba's long association with the United Nations began in 1972 when he led the Egyptian delegation to the Stockholm Conference on the Human Environment, a meeting at which the United Nations Environment Programme (UNEP) was established. A year later, he was appointed deputy executive director of UNEP and in 1976 executive director of the organization, a post he still holds.

Tolba has published extensively both in the field of science (about 95 papers) and in the general field of environmental sciences (more than 200 statements and articles). He has been awarded honorary degrees from Moscow University (1978), Korea's Hanyang University (1987), and Kenyatta University (1989). In addition, he has been awarded the Environment Medal from the government of Argentina (1983), the Ramdco Medal from the government of India (1984), the Environmental Memorial Medal Award from the government of Czechoslovakia (1988), and the Environmental Medal of the Better World Society (1988).

Henry A. Waxman (1939–)

Henry Waxman was born on 12 September 1939, in Los Angeles. He attended the University of California at Los Angeles, from which he received his bachelor's degree in political science in 1961 and his J.D. from the Law School in 1964. After working as an attorney for four years, Waxman won a seat in the California Assembly, where he served for six years.

In 1974 Waxman ran for Congress in California's 29th District, a district that includes the cities of Hollywood, West Hollywood, Santa Monica, Beverly Hills, and parts of Los Angeles, Brentwood, Westwood, and other Los Angeles suburbs. He won the election and rapidly earned a reputation in Congress for his knowledge of and interest in the fields of health and the environment. In 1978 he was chosen to be chairman of the Subcommittee on Health and the Environment of the Committee on Energy and Commerce. His selection was unusual in that it was one of the few times in history when the majority party had chosen someone for a chairmanship for reasons other than seniority.

In his post on Health and the Environment, Waxman has devoted his attention, among other issues, to the question of ozone depletion. He held hearings in the 1980s on this issue and worked diligently to prevent the Reagan administration from changing the Clean Air Act in 1981 and 1982. He then played a major role in developing provisions of the 1990 amendments to the Clean Air Act, amendments that included provision for protection of the ozone layer.

Waxman has won reelection to his House seat in every election since 1974. In addition to his work on environmental issues, Waxman has championed health care reform and was one of the leaders in efforts to pass President Bill Clinton's 1994 health reform package. He has also taken a special interest in Middle Eastern affairs and in efforts to guarantee the security and safety of Israel.

Data, Opinions, and Documents

This chapter provides three types of information relevant to the debate over ozone depletion: data, opinions, and documents. The first of these sections presents factual information concerning the loss of ozone from the stratosphere over the Antarctic and worldwide, increases in ultraviolet levels at the Earth's surface, characteristics of ozone-depleting chemicals, and similar data. The second section presents a broad range of opinions about ozone depletion. Included in this section are statements from those who are convinced of the existence of an ozone problem and who argue for strong action to deal with that problem, from those who doubt the existence of such a problem and insist that no action should be taken, and from those who hold some intermediate position and argue for a moderate approach. The third section offers a selection of treaties, agreements, laws, regulations, and other actions taken as a way to deal with stratospheric ozone depletion.

Data

Trends in Antarctic Ozone Levels

The graph below plots chlorine monoxide and ozone concentrations against latitude. This graph is often referred to as the "smoking gun" that shows the connection between CFCs and ozone depletion. Note two features of the graph: First, the farther south one goes (the closer to the South Pole), the lower the concentration of ozone and the higher the concentration of chlorine monoxide. Chlorine monoxide is the chemical implicated in the loss of ozone. You can see that the ozone "hole" begins at about 69° south latitude. Second, as the concentration of chlorine monoxide increases,

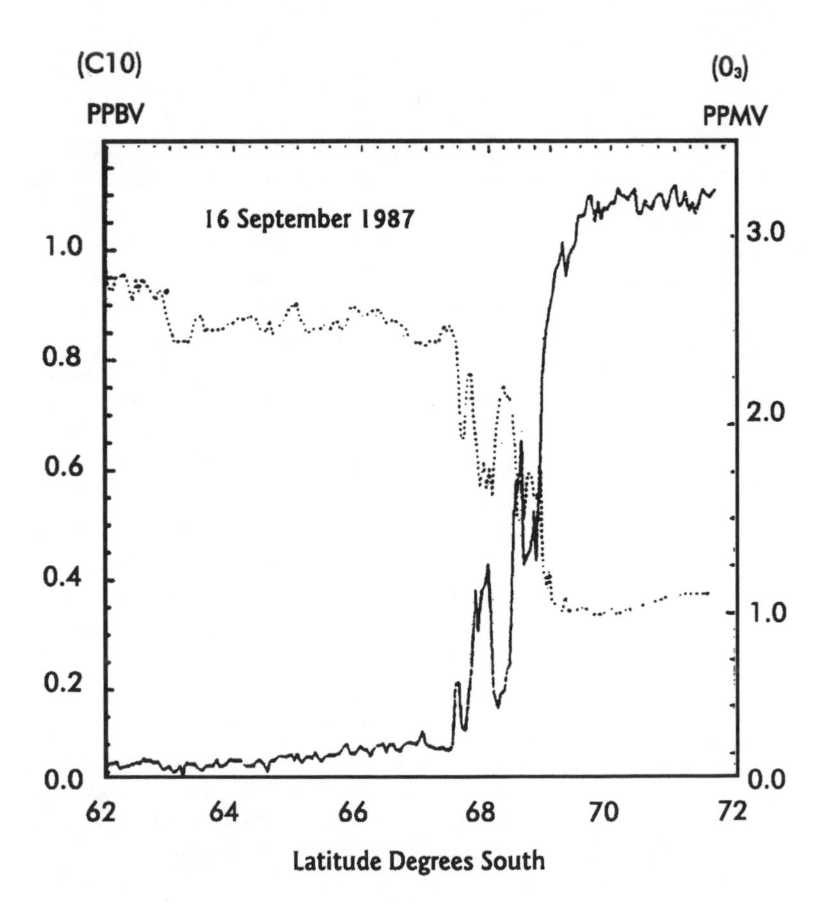

the concentration of ozone decreases. That relationship would suggest that the presence of chlorine monoxide (a product of the decomposition of CFCs) is associated with the loss of ozone.

Trends in Worldwide Ozone Concentrations

Table 1 summarizes changes in average total ozone abundances as measured at 24 individual monitoring stations during the 22-year period between 1965 and 1986, inclusively. The numbers given show the increase (+) or decrease (–) in ozone abundance for the period 1976–1986 compared to the period 1965–1975. The second term in each measurement (±) indicates the range of error in the measurement. Table 2 shows the same data organized by latitude.

TABLE 1 Comparison of Ozone Abundances between 1976–1986 and 1965–1975, as Recorded at 24 Monitoring Stations

Latitude	Station	Winter	Summer	Annual
74.7°N	Resolute (Canada)	-1.4 ± 1.8	-0.8 ± 0.9	-1.6 ± 1.0
64.1°N	Reykjavik (Iceland)	-2.5 ± 2.2	+1.7 ± 1.3	+0.1 ± 2.4
60.2°N	Lerwick (Scotland)	-3.8 ± 2.0	-0.9 ± 0.9	-1.6 ± 1.0
58.8°N	Churchill (Canada)	-4.2 ± 0.9	-1.4 ± 0.8	-2.5 ± 0.7
53.6°N	Edmonton (Canada)	-4.7 ± 1.3	+0.8 ± 0.9	-1.8 ± 0.8
53.3°N	Goose Bay (Canada)	-2.4 ± 1.3	-0.1 ± 1.1	-0.8 ± 0.9
51.8°N	Belsk (Poland)	-3.2 ± 0.8	-1.2 ± 1.0	-1.2 ± 0.9
50.2°N	Hradec Kralove (Czech Republic)	-4.7 ± 2.0	+1.1 ± 0.9	-1.8 ± 1.1
47.8°N	Hohenpeissenberg (Germany)	-1.8 ± 1.7	+0.2 ± 0.9	-1.0 ± 0.9
46.9°N	Caribou (Maine/US)	-2.8 ± 1.5	-0.6 ± 0.8	-1.8 ± 0.9
46.8°N	Arosa (Switzerland)	-3.0 ± 1.3	-1.1 ± 1.0	-2.0 ± 0.9
46.8°N	Bismarck (ND/US)	-3.0 ± 1.2	-1.4 ± 1.0	-2.0 ± 0.7
43.8°N	Toronto (Canada)	-1.3 ± 1.2	-1.3 ± 0.8	-1.2 ± 0.7
43.1°N	Sapporo (Japan)	-0.6 ± 1.4	-0.1 ± 0.9	-0.3 ± 0.6
42.1°N	Vigne di Valle (Italy)	-2.9 ± 1.2	+0.7 ± 0.9	-0.9 ± 0.9
40.0°N	Boulder (CO/US)	-3.9 ± 1.3	-3.1 ± 0.7	-3.3 ± 0.8
39.3°N	Cagliari (Italy)	-2.5 ± 1.7	-0.7 ± 1.1	-1.1 ± 1.2
36.3°N	Nashville (TN/US)	-1.8 ± 1.4	-3.3 ± 0.7	-2.4 ± 0.8
36.1°N	Tateno (Japan)	-0.7 ± 1.6	-0.5 ± 0.8	-0.4 ± 0.7
31.6°N	Kagoshima (Japan)	+0.9 ± 1.7	+0.5 ± 1.0	+0.9 ± 0.8
30.4°N	Tallahassee (FL/US)	-1.7 ± 1.9	-0.2 ± 1.1	-0.7 ± 0.8
30.2°N	Quetta (Pakistan)	-1.1 ± 1.6	+0.1 ± 0.8	-0.7 ± 0.8
25.5°N	Varanasi (India)	-0.3 ± 1.4	+0.4 ± 0.9	-0.2 ± 0.9
19.5°N	Mauna Loa (HI/US)	-1.5 ± 1.7	0.0 ± 0.6	-0.9 ± 0.6

Source: World Meteorological Organization, *Scientific Assessment of Stratospheric Ozone: 1989* (Geneva: World Meteorological Organization, 1990), p. 21.

TABLE 2 Comparison of Ozone Abundances by Latitude between 1976–1986 and 1965–1975

North Latitude	Winter	Summer	Annual
30°N to 39°N	-1.2 ± 1.5	-0.7 ± 1.0	-0.8 ± 1.1
40°N to 60°N	-3.0 ± 0.9	-0.4 ± 0.5	-1.6 ± 0.6
30°N to 60°N	-2.5 ± 1.5	-0.5 ± 0.6	-1.4 ± 0.7

Source: World Meteorological Organization, *Scientific Assessment of Stratospheric Ozone: 1989* (Geneva: World Meteorological Organization, 1990), p. 21.

Chlorofluorocarbons and Related Compounds

The chemical substances most commonly associated with ozone depletion are chlorofluorocarbons (CFCs), fluorocarbons (halons), and related compounds. Table 3 lists some of the most common of these compounds along with their chemical structures, formulas, common name(s), and physical state at room temperature.

TABLE 3 Chlorofluorocarbons and Related Compounds

Chemical Name	Formula	Common Name(s)	Physical State (at room temperature)
trichlorofluoromethane	CCl_3F	CFC-11, Freon-11	liquid (boils at 23.7°C)
dichlorodifluoromethane	CCl_2F_2	CFC-12, Freon-12	gas
chlorotrifluoromethane	$CClF_3$	CFC-13, F-13	gas
1,2-difluoro-1,1,2,2-tetrachloroethane	CCl_2FCCl_2F	CFC-112, F-112	solid
1,1-difluoro-1,2,2,2-tetrachloroethane	$CClF_2CCl_3$	CFC-112a, F-112a	solid
1,1,2-trichloro-1,2,2-trifluoroethane	CCl_2FCClF_2	CFC-113, Freon-113	liquid
1,1,1-trichloro-2,2,2-trifluoroethane	CCl_3CF_3	CFC-113a	liquid
1,2-dichloro-1,1,2,2-tetrafluoroethane	$CClF_2CCl$	CFC-114, F-114	gas
1,1-dichloro-1,2,2,2-tetrafluoroethane	CCl_2FCF_3	CFC-114a, F-114a	gas

Table 3, *continued*

Chemical Name	Formula	Common Name(s)	Physical State (at room temperature)
1-chloro-1,1,2,2,2-pentafluoroethane	$CClF_2CF_3$	CFC-115, F-115	gas
bromochlorodifluoromethane	CF_2BrCl	halon 1211	gas
bromotrifluoromethane	CF_3Br	halon 1301	gas
dibromotetrafluoroethane	$C_2F_4Br_2$	halon 2402	gas
trichloromethane	$CHCl_3$	chloroform	liquid
tetrachloromethane tetrachloride	CCl_4	carbon tetrachloride	liquid
bromomethane	CH_3Br	methyl bromide	liquid
1,1,1-trichloroethane	$C_2H_3Cl_3$	methyl chloroform	liquid

Production and Release of Chlorofluorocarbons, 1935–1990

Table 4 summarizes the production and release of the two most common CFCs, CFC-11 and CFC-12, from 1935 to 1992.

TABLE 4 CFC Production and Emission Rates, 1935–1992 (in million kilograms)

Year	Production	Estimated Release
1935	1.0	0.3
1945	20.5	8.3
1955	83.9	71.2
1965	312.9	283.5
1975	695.1	715.0
1985	703.1	649.3
1990	463.9	526.6
1992	402.6	426.4

Source: Alternative Fluorocarbons Environmental Acceptability Study, *Production and Atmospheric Release Data for CFC-11 and CFC-12 (through 1992)* (Washington, DC: AFEAS Administrative Organization, September 1993).

Estimated Use Patterns of CFC-11 and CFC-12 in 1985

Table 5 shows the ways in which the two most common CFCs, CFC-11 and CFC-12, were used in the United States and the rest of the world in a peak year, 1985. The figures indicate the percentage of the substance that was utilized for the use cited.

TABLE 5 U.S. and World Uses of CFC-11 and CFC-12 in 1985

	CFC-11		CFC-12	
Use	U.S.	World	U.S.	World
Blowing agent	71%	58%	11%	12%
Aerosols	5%	31%	4%	32%
Refrigeration	6%	3%	(see subcategories below)	
Mobile air-conditioning	(included above)		37%	20%
Commercial food refrigeration	(included above)		4%	3%
Home refrigerators	(included above)		2%	3%
Other cooling systems	(included above)		1%	1%
Miscellaneous	18%	8%	10%	7%
Unallocated	—	—	31%	22%

Sources: Various sources as cited in *Fully Halogenated Chlorofluorocarbons* (Geneva: World Health Organization, 1990), Tables 6 and 7, pp. 36–37, and Cynthia Pollack Shea, *Protecting Life on Earth: Steps to Save the Ozone Layer,* Paper 87 (Washington, DC: Worldwatch Institute, 1988), Table 5, p. 27.

Trends in U.S. CFC Use, 1976–1992

The dramatic changes in the amounts of CFC used in the United States between 1976 and 1992 are illustrated in Table 6. Note in particular the large change in the amount of CFCs used in aerosols after the 1978 ban by the Environmental Protection Agency.

TABLE 6 Use by Type in the United States, 1975–1990

	Refrigeration	Blowing Agent	Aerosol Propellant	All Other Uses
1976	154,721	111,948	432,275	51,617
1978	186,699	145,785	307,219	41,231
1980	191,280	173,318	243,217	32,023
1982	180,134	176,902	208,471	33,975
1984	211,439	223,148	218,834	41,041
1986	224,143	253,353	224,076	46,939
1988	267,856	312,024	176,550	40,558
1990	181,912	210,790	48,694	22,470
1992	195,178	169,745	22,866	14,852

Source: Alternative Fluorocarbons Environmental Acceptability Study, *Total CFC-11/12 Sales by Category for the Years 1976 through 1992 from Reporting Companies,* Schedules 5 and 6 (Washington, DC: AFEAS Administrative Organization, September 1993).

Ozone Depletion Potentials and Global Warming Potentials of Certain CFCs and Related Compounds

Scientists have calculated the relative amount of ozone destroyed by a number of CFCs and related compounds and their relative contributions to a predicted rise in atmospheric temperature. These measures are known as Ozone Depletion Potential (ODP) and Global Warming Potential (GWP). The higher the figure, the greater the threat. CFC-11 is taken as the standard for both measurements. The values in Table 7 are averages from more than one model.

TABLE 7 Ozone Depletion and Global Warming Potentials of Selected CFCs and Related Compounds

Compound	ODP	GWP
CFC-11	1.0	1.0
CFC-12	0.9	3.1
CFC-113	0.9	1.3
CFC-114	0.7	3.9
CFC-115	0.4	7.5
HCFC-22	0.05	0.34
HCFC-123	0.017	0.018
HCFC-124	0.020	0.096
HFC-125	0	0.58
HFC-134a	0	0.26
HCFC-141b	0.09	0.090
HCFC-142b	0.05	0.36
HCFC-143a	0	0.74
HFC-152a	0	0.030
CCl_4	1.1	0.34
CH_3CCL_3	0.12	0.024

Source: Adapted from World Meteorological Organization, *Scientific Assessment of Stratospheric Ozone: 1989* (Geneva: World Meteorological Organization, 1990), p. xiii.

Chlorine Loading Potential for Various Gases

The contribution of any particular gas to ozone depletion depends on a number of factors. Two of the most important are the length of time the gas is expected to remain in the atmosphere and the number of chlorine atoms released per molecule of the gas. Table 8 shows the approximate expected lifetime for each gas and its Chlorine Loading Potential (CLP), which is the maximum number of chlorine atoms contributed to the stratosphere compared to an equal weight of CFC-11.

TABLE 8 Lifetimes and Chlorine Loading Potentials of Selected CFCs and Related Compounds

Compound	Lifetime (in years)	CLP
CFC-11	60.0	1.0
CFC-12	120.0	1.5
CFC-113	90.0	1.11
CFC-114	200.0	1.8
CFC-115	400.0	2.0
HCFC-22	15.3	0.14
HCFC-123	1.6	0.016
HCFC-124	6.6	0.04
HFC-125	28.1	0
HFC-134a	15.5	0
HCFC-141b	7.8	0.10
HCFC-142b	19.1	0.14
HCFC-143a	41.0	0
HFC-152a	1.7	0
CCl_4	50.0	1.0
CH_3CCl_3	6.3	0.11

Source: Adapted from World Meteorological Organization, *Scientific Assessment of Stratospheric Ozone: 1989* (Geneva: World Meteorological Organization, 1990), p. 432.

Variations in Ultraviolet Levels

One point made by those who doubt the reality or seriousness of ozone depletion has been that no concrete evidence exists for an increase in levels of ultraviolet radiation reaching the Earth that would be expected as a result of lower ozone concentrations in the stratosphere. In November 1993 two scientists at Environment Canada, J. B. Kerr and C. T. McElroy, reported just such evidence. The abstract of their article, reprinted below, confirms that in at least one location, levels of UV radiation have, indeed, increased over the last half decade.

> Spectral measurements of ultraviolet-B radiation made at Toronto since 1989 indicate that the intensity of light at wavelengths near 300 nanometers has increased by 35 percent per year in winter and 7 percent per year in summer. The wavelength dependence of these trends indicates that the increase is caused by the downward trend in total ozone that was measured at Toronto during the same period. The trend at wavelengths between 320 and 325 nanometers [a range at which ozone does *not* absorb radiation] is essentially zero.

Source: J. B. Kerr and C. T. McElroy, "Evidence for Large Upward Trends of Ultraviolet-B Radiation Linked to Ozone Depletion," *Science*, 12 November 1993, p. 1032.

Scientific Assessment of Stratospheric Ozone: 1989

In 1988, the World Meteorological Organization, under provisions of the 1987 Montreal Protocol, sponsored two meetings on Scientific Assessment of Stratospheric Ozone: 1989. Those meetings were held at The Hague, The Netherlands, on 17–18 October, and in London on 29 November. The following are the primary conclusions that were reached at these meetings.

Polar Ozone

- There has been a large, rapid, and unexpected decrease in the abundance of springtime Antarctic ozone over the last decade.
- The weight of scientific evidence strongly indicates that man-made chlorine and bromine compounds are primarily responsible for the ozone loss in Antarctica.
- While the onset of the Antarctic ozone hole is linked to the recent growth in the atmospheric abundance of chlorofluorocarbons (CFCs) and to a lesser extent bromine compounds, many of its features are influenced by meteorological conditions.
- The chemical composition of the Arctic stratosphere was found to be highly perturbed.

Global Trends

- Several recent analyses of total column ozone data support the conclusion of the 1988 International Ozone Trends Panel (TOP) that there is a downward trend in ozone during winter at mid-to-high latitudes in the Northern Hemisphere over the past two decades.
- Substantial uncertainties remain in defining changes in the vertical distribution of ozone.
- Recent measurements suggest that the rate of growth in atmospheric methane has slowed somewhat.

Theoretical Predictions

- Theoretical models do explain many of the general features of the atmosphere, but new limitations have been recognized.
- Current understanding predicts that, if substantial emissions of halocarbons continue, the atmospheric abundances of chlorine and bromine will increase and, as a result, significant ozone decreases, even outside of Antarctica, are highly likely in the future.

Halocarbon Ozone Depletion and Global Warming Potentials (ODPs and GWPs)

- The impact on stratospheric ozone of the halocarbons (HCFCs and HFCs) that are proposed as substitutes for the CFCs depends upon their chemical removal processes in the lower atmosphere (troposphere).
- The values of the ODPs for the HCFCs are significantly lower than those for the CFCs.
- The steady-state values of the halocarbon GWPs of the HCFCs and HFCs are lower than those of the CFCs.

Source: World Meteorological Organization, *Scientific Assessment of Stratospheric Ozone: 1989* (Geneva: World Meteorological Organization, 1990), pp. vii–xiii.

First Announcement of the Ozone "Hole" over the Antarctic

One of the key events in the story of ozone depletion over the past few decades has been the report issued by J. C. Farman and his colleagues from the British Antarctic Survey in the 16 May 1985 issue of the scientific journal *Nature.* The abstract of that article is reprinted below. Essentially, the authors say that, while models have been predicting loss of ozone from the stratosphere, the actual observations made at Halley Bay greatly exceed any estimates made thus far. Asterisks (*) indicate that footnotes in the original article have been deleted in this reprint.

> Recent attempts* to consolidate assessments of the effect of human activities on stratospheric ozone (O_3) using one-dimensional models for 30°N have suggested that perturbations of total

> O_3 will remain small for at least the next decade. Results from such models are often accepted by default as global estimates.
>
> *The inadequacy of this approach is here made evident by observations that the spring values of total O_3 in Antarctica have now fallen considerably. The circulation in the lower stratosphere is apparently unchanged, and possible chemical causes must be considered. We suggest that the very low temperatures which prevail from midwinter until several weeks after the spring equinox make the Antarctic stratosphere uniquely sensitive to growth of inorganic chlorine, ClX [*sic*], primarily by the effect of this growth on the NO_2/NO ratio. This, with the height distribution of UV irradiation peculiar to the polar stratosphere, could account for the O_3 losses observed.

Source: J. C. Farman, B. G. Gardiner & J. D. Shanklin, "Large Losses of Total Ozone in Antarctica Reveal Seasonal ClO_x/NO_x interaction," *Nature*, 16 May 1985, p. 207.

Reduction in Atmospheric Halons

Some encouraging news about ozone-depleting chemicals was announced in October 1992 by scientists at the National Oceanic and Atmospheric Administration's Climate Monitoring and Diagnostics Laboratory and the University of Colorado's Cooperative Institute for Research in Environmental Sciences. The research team reported in the journal *Nature* that atmospheric levels of certain halons had already begun to diminish. These results seemed to bode well for efforts to reduce ozone depletion. The abstract of this article is reprinted below.

> Halons H-1301 ($CBrF_3$) and H-1211 ($CBrClF_2$) have been introduced into the atmosphere, mainly through use in fire extinguishers, for almost three decades. Although each is now present in the troposphere at a concentration of only 2 parts per 10^{12}, these gases have long atmospheric lifetimes . . . and carry significant amounts of bromine to the stratosphere, where it can destroy ozone catalytically. For this reason, the halons have high ozone depletion potential. The manufacture of both gases is to be discontinued globally by the year 2000, according to the Montreal Protocol, and perhaps sooner, as a result of unilateral action by users, manufacturers and producing countries. Here we present a six-year record of tropospheric halon mixing ratios which shows that the growth rates of H-1301 and H-1211 have already begun to decrease substantially. This recent decrease in growth rates is consistent with industry emission estimates (although these have greater uncertainties), and supports current appraisals of atmospheric lifetimes.

> Our results suggest that, even though these halons are relatively long-lived species, their atmospheric mixing ratios may stabilize or begin to decrease within the next few years.

Source: J. H. Butler, J. W. Elkins, B. D. Hall, S. O. Cummings, and S. A. Montzka, "A Decrease in the Growth Rates of Atmospheric Halon Concentrations," *Nature*, 1 October 1992, p. 403.

Projections for Ozone Depleting Chemicals

What is the prognosis for the fate of ozone-depleting chemicals in the atmosphere over the next few centuries? Two of the leading researchers on ozone depletion attempted to answer that question in a 1992 article published in the journal *Nature*. Essentially, the researchers argued that CFCs are likely to remain a threat to the ozone layer for many years into the future. On the other hand, their most likely substitutes, the HCFCs and HFCs, are likely to pose a short-term threat to the ozone layer, but will be of much less concern after about a decade. A major conclusion from this study is that HCFCs and HFCs may be good alternatives to CFCs in the long term, but they are still a significant threat to ozone in the near term. The numbers in Table 9 represent ozone depletion potentials for a sample of ozone-depleting chemicals, all compared to CFC-11 as a standard equal to 1.0. The data reported here are a combination of observed and predicted behaviors.

TABLE 9 Ozone Depletion Potentials of Selected Chemical Compounds Relative to CFC-11

	Years into the Future								
Substances	5	10	15	20	25	30	40	100	500
CFC-11	1.0	1.0	1.0	1.0	1.0	1.0	1.0	1.0	1.0
CFC-113	0.55	0.56	0.58	0.59	0.60	0.62	0.64	0.78	1.09
CCl4	1.26	1.25	1.24	1.23	1.22	1.22	1.20	1.14	1.08
CH_3Br	15.3	5.4	3.1	2.3	1.8	1.5	1.2	0.69	0.57
HCFC-142b	0.17	0.16	0.15	0.14	0.13	0.13	0.12	0.08	0.07
HCFC-141b	0.54	0.45	0.38	0.33	0.30	0.26	0.22	0.13	0.11
HCFC-123	0.51	0.19	0.11	0.08	0.07	0.06	0.04	0.03	0.02
HCFC-124	0.17	0.12	0.10	0.08	0.07	0.06	0.05	0.03	0.02
H-1211	11.3	10.5	9.7	9.0	8.50	8.0	7.1	4.9	4.1
H-1301	10.3	10.4	10.5	10.5	10.6	10.7	10.8	11.5	12.5

Source: Adapted from Susan Solomon and Daniel Albritton, "Time-Dependent Ozone Depletion Potentials for Short- and Long-Term Forecasts," *Nature*, 7 May 1992, p. 35.

Opinions

Article by Rowland and Molina

One might argue that the symbolic beginning of the current debate over stratospheric ozone depletion actually began in 1974 with the publication of an article by F. Sherwood Rowland and Mario Molina. Rowland and Molina put forth the hypothesis in this article that CFCs might be a significant cause for the destruction of ozone in the stratosphere. The brief abstract of that article, reprinted below, provides a succinct statement of their viewpoint.

> Chlorofluoromethanes are being added to the environment in steadily increasing amounts. These compounds are chemically inert and may remain in the atmosphere for 40–150 years, and concentrations can be expected to reach 10–30 times their present levels. Photodissociation of the chlorofluoromethanes in the stratosphere produces significant amounts of chlorine atoms, and leads to the destruction of stratospheric ozone.

Source: Molina, M. J., and F. S. Rowland, "Stratospheric Sink for Chlorofluoromethanes: Chlorine-Atom Catalyzed Destruction of Ozone," *Nature*, 28 June 1974, p. 810.

The publication of the Molina-Rowland paper set off an intense debate about the depletion of stratospheric ozone. That debate took place on a number of levels and included disagreements as to the scientific validity of the Molina-Rowland theory, the likelihood of its being a significant issue in the real world, and actions, if any, that should be taken to deal with the issue. The following excerpts illustrate some of the debate that was going on shortly after 1974 in the professional and general literature.

"Earth's Sunscreen Is Rotting Away"

In its January 1975 issue, the magazine of popular science, *Science Digest*, announced the Rowland-Molina theory of ozone destruction under the above heading. In an accompanying photo caption, the magazine warned that "the combined effect of billions of aerosol cans releasing their chemical propellant into the atmosphere is gradually eroding the ozone layer." In the article, the claim was made that

> Ozone, that fragile shell of gas that helps shield the earth from lethal doses of ultraviolet solar radiation, may be the latest on the list of victims of man's carelessness. The cause of this destruction is the innocuous spray can. . . .
>
> Since the earth has never been stripped of this sun filter, it is impossible to say definitely what would happen with it gone. According to some scientific guesswork, more incoming ultraviolet radiation could heat up the atmosphere, making global climate changes, and could also destroy or damage plants or microscopic organisms valuable for producing oxygen.

Source: "Earth's Sunscreen Is Rotting Away," *Science Digest*, January 1975, pp. 18–19.

Statement by S. Robert Orfeo, Allied Chemical, 1975

Not everyone was as concerned about the ozone layer as were the editors of *Science Digest*. This was particularly true of those associated with the production and sale of CFCs. The following statement, by the director of applied research in Allied Chemicals Specialty Chemicals Division, is typical of those individuals and groups. The argument offered in most cases was that sufficient information was not yet available on which to start making policies that would include restrictions on or the banning of CFCs.

> The validity of the Rowland-Molina hypothesis has not been established. There is no concrete evidence to show that the ozone-depleting reaction with chlorine in fact takes place in the stratosphere. As a matter of fact, detailed analysis of the available ozone data indicates that the ozone level in the stratosphere actually increased in the 1960's, a period of high production of chlorine and chlorine containing products.
>
> Furthermore, continued monitoring of the atmosphere has revealed the presence of many chlorine containing compounds in the troposphere and lower stratosphere. Therefore, if there is a problem, it should be considered as a potential halogen problem and not restricted to halocarbons.
>
> Also, the fluctuations in ozone levels observed in recent years seem to be merely a continuation of historical cycles. If such historical patterns persist, as they have through 1974 based on the Arosa, Switzerland data, an upward trend should be discernable in the late 1970's after passing through a minimum in the mid-1970's.
>
> As to whether there is time to complete the studies aimed at resolving the ozone depletion issue, it should be noted that the

most recent forecasts based on the latest reaction rate chemistry have resulted in a downward revision of the short term impact of chlorine in the ozone level. The updated model predicts that continued production for the next three years, even allowing for a 10% annual growth rate, followed by a halt would result in a slightly higher depletion.

If production is stopped now, the models project a 1.2% depletion by 1990. If production is allowed to continue for three years, the depletion according to the models would be only 1.7% maximum in 1990.

Source: "Allied's Orfeo Stresses Uncertainties in Ozone Theory," *Aerosol Age*, June 1975, pp. 62–63.

A Neutral Observer's View of the Issue, 1978

The ever-shifting debate over ozone depletion appeared to have calmed somewhat in 1978, at least in the eyes of one science reporter. The following extract comes from a report on "Fluostrat '78," a symposium sponsored by the Society of Chemical Industry in Brighton, England.

The panic about the supposed threat to the ozone layer from chlorofluorocarbons (CFCs) is over, and has been replaced by reasoned scientific debate. . . .

In 1976, "scare" stories suggested that these very long lived compounds might be transferring chlorine to the stratosphere, where it would join in photochemical reactions breaking down ozone. This would permit ultraviolet radiation, normally blocked by ozone in the stratosphere, to reach the ground. The scare was led by the known association of ultraviolet radiation and cancer, giving doomsters a platform from which to preach about the destructive influence of mankind's thoughtless use of technology on the environment. But as more attention has been concentrated on the many facets of this "ozone problem," the great complexity of the systems by which the atmosphere is kept in balance has become apparent. Spokesmen for the CFC industry emphasised in Brighton the strength of the research effort which they are financing, and have been running since 1972, to determine the ultimate fate of these compounds. They reiterated that, in their view, the evidence to date is insufficient to justify a panic ban on the use of CFCs F_{11} [CFC-11] and F_{12} [CFC-12] (the two at the heart of the debate). The scientists contributing to the symposium backed the message up in no uncertain manner.

Source: John Gribbin, "Ozone Passion Cooled by the Breath of Sweet Reason," *New Scientist*, 12 October 1978, p. 94.

Environmental Protection Agency Statement, 1980

By 1980, at least some scientists and government officials had become convinced that ozone depletion was a real problem that had to be addressed by regulations. The Environmental Protection Agency announced its intention to limit production of chlorofluorocarbons as a first step in dealing with this problem. The formal announcement of that policy in a statement by Barbara Blum, Deputy Administrator of the Environmental Protection Agency at the International Meeting on Chlorofluorocarbons in Oslo, Norway, on 15 April 1980 (subsequently read into the record at a congressional hearing), is reprinted below.

> I am today announcing that the United States will propose limiting the production of ozone-destroying chlorofluorocarbons to our 1979 production level. We expect that this action will hold U.S. production to about 551 million pounds per year. . . .
>
> We in the United States are taking this step because of continuing studies showing that worldwide chlorofluorocarbon emissions jeopardize public health and the environment. For example, the most recent report by the U.S. National Academy of Sciences states that continued global emissions of chlorofluorocarbons, even if restricted to the 1977 level, most likely will lead to a 16 percent reduction in the earth's stratospheric ozone. Obviously, any growth above total world 1977 use levels will lead to greater ozone depletion. This, in turn, would create a 44 percent increase in the amount of harmful ultraviolet radiation reaching the earth's surface. Such radiation would lead to thousands of additional cases of human melanoma skin cancer — which often is fatal — and hundreds of thousands of additional cases of the non-melanoma variety. This radiation would also cut production of wheat, corn, soybeans, and rice and of commercially important fish such as anchovies, mackerel, shrimp, and crab. . . .
>
> The action I am announcing today conveys the urgent and deep concern of the United States about the threat chlorofluorocarbons continue to pose. Our country is moving forward now because we believe that chlorofluorocarbons comprise one of the leading international environmental issues of the decade.

Source: "EPA Proposed Rulemaking on Chlorofluorocarbons (CFCs) and Its Impact on Small Business," Hearings before the U.S. House of Representatives, Committee on Small Business, Subcommittee on Antitrust and Restraint of Trade Activities Affecting Small Business, July 15, 1981, pp. 58–59.

Industry Response to EPA Plans To Limit CFC Production, 1981

Industries for whom CFCs were a critical product were not willing in 1981 to accept the need for regulation of these important chemicals. At the same hearings at which Ms. Blum's statement was read into the record, the Alliance for a Responsible CFC Policy, an industry group, presented the following testimony.

> The Alliance strongly opposes any further regulation of CFCs at this time. Such regulation cannot be justified in light of (a) the lack of any evidence that ozone is being depleted as hypothesized, (b) the availability of an early warning system, (c) the substantial uncertainties and lack of international agreement with respect to the science concerning both the ozone depletion hypothesis and the possible effects (if any) of any ozone depletion upon man and the environment, (d) the insignificant risk attributable to a delay in regulation, and (e) the tremendous public benefits which depend upon the continued ability to purchase and use CFCs at satisfactory levels and affordable prices.
>
> EPA should not regulate on the basis of unverified scientific theory. Instead, the Alliance urges EPA to defer any further regulation until such time as there is an international consensus regarding the science and the need for regulatory action. In the interim, the statistical technique of time-trend analysis, applied to actual measurements of ozone concentration, will provide an "early warning system" capable of detecting even small changes in the stratospheric ozone concentration. . . .
>
> The Alliance concludes that:
>
> - Ozone trend analysis provides an early warning system for any developing problem in stratospheric ozone.
> - Ozone depletion has *not* been measured.
> - *Calculated* ozone depletion is about one-half or less of that stated in the NAS Report.
> - Preliminary results, if confirmed, will substantially reduce ozone depletion estimates even further.
> - Stratospheric science is in a period of very rapid change.
> - The current status of the ozone depletion theory does not justify further regulation of CFCs.
> - There is negligible risk in waiting for the results from further research.

Source: "EPA Proposed Rulemaking on Chlorofluorocarbons (CFCs) and Its Impact on Small Business," Hearings before the U.S. House of Representatives, Committee on Small Business, Subcommittee on Antitrust and Restraint of Trade Activities Affecting Small Business, 15 July 1981, pp. 76, 106–107.

The announcement by Joseph Farman and his colleagues at Halley Bay in 1985 of a "hole" in the ozone layer reinvigorated the ongoing debate over ozone depletion. An important forum for that debate was the hearing rooms of the U.S. Congress, where authorities on both sides of the dispute presented their cases before a variety of Senate and House subcommittees. The selections below provide a flavor of the kinds of arguments being made about the possibility of ozone depletion and the kinds of actions that should be taken about it.

Statement of Robert T. Watson, National Aeronautics and Space Administration, 1987

> In my opinion, protection of the ozone layer is one of the most important global environmental issues of our time. . . . Based on our current understanding, it is clear that an uncontrolled expansion of the production and subsequent emission of chlorofluorocarbons into the Earth's atmosphere would lead to significant reduction of the total column concentration of ozone during the next several decades which could lead to adverse consequences for human health and aquatic and terrestrial ecosystems. In addition there would be a substantial redistribution of the vertical distribution of ozone which could modify the climate system. Consequently, even in the face of scientific uncertainties, prudence would dictate that it is urgent for us to devise at the international level a short term, as well as a long term, strategy to limit the emission of these chemicals into the atmosphere.

Source: "Ozone Layer Depletion," Hearings before the U.S. House of Representatives, Committee on Energy and Commerce, Subcommittee on Health and the Environment, 9 March 1987, p. 61.

EPA's Risk Assessment on Stratospheric Ozone Modification, Testimony by Lee M. Thomas, 1987

> To provide a sound scientific basis for future decisions, EPA staff prepared a five-volume risk assessment summarizing and synthesizing the current scientific understanding of this issue. . . .
>
> I would like to recount what I believe to be the most important findings from the broad range of issues covered by the risk assessment:
>
> - Atmospheric processes affecting the creation and destruction of ozone are extremely complex. While our knowledge of them

has substantially increased over the past decade, we have much more to learn.

- A consensus appears to exist that if enough chlorine from CFC's or other substances is put into the atmosphere and subsequently into the stratosphere, ozone depletion is likely to occur.
- Because CFC's have atmospheric lifetimes of 75–120 years, emissions today will influence stratospheric ozone concentrations for a very long period of time.
- If we were to wait for ozone depletion to occur before reducing CFC emissions, any depletion would continue to worsen for several decades.
- While increases in some gases (e.g., carbon dioxide and methane) may, in part, buffer the loss or depletion due to CFC's, all these gases appear to cumulatively add to the greenhouse effect and to global warming. . . .
- If ozone depletion occurs, the resulting increases in ultraviolet radiation (UV-B) could pose substantial risks to human health, to a wide range of other organisms and to the environment. . . .
- The exact magnitude of the impacts noted above is difficult to quantify and largely depends on the rate of growth of emissions of CFC's and other gases. . . .

In general, the risk assessment indicates that the potential for substantial health and environmental risks exists, but that many scientific questions remain to be answered.

Source: "Ozone Layer Depletion," Hearings before the U.S. House of Representatives, Committee on Energy and Commerce, Subcommittee on Health and the Environment, 9 March 1987, pp. 95–96.

The United States' Position (in 1986) Regarding International Regulation of CFCs

The United States believes that the potential risks to the stratospheric ozone layer from certain man-made chemicals require early and concerted action by the international community. Since the adoption in Vienna in March 1985 of the Ozone Layer Convention, an intensive scientific research and technical analysis effort has been carried out and is continuing, as reflected in the recent series of UNEP [United Nations Environment Programme]–sponsored workshops. The results continue to indicate the emergence of a serious environmental problem of global proportions.

The United States further believes that governments should pursue three broad objectives during the course of the negotiations, to be embodied and elaborated in the final protocol. These are:

A. Agreement on a meaningful near-term first step to reduce significantly the risk of stratospheric ozone depletion and associated environmental and human health impacts.
B. Agreement on a long-term strategy and goals for coping with the problem successfully.
C. Agreement on a carefully-scheduled plan for achieving the long-term goals, including periodic reassessment and appropriate modification of the strategy and goals in response to new scientific and economic information.

In response to UNEP's invitation, the U.S. has prepared for discussion purposes a draft text based on the U.S. views statement which we recently circulated.

[The U.S. draft text then lists five more articles dealing with control measures; calculation of aggregate annual emissions; assessment and adjustment of control measures; control of trade; and reporting of information.]

Source: "Ozone Layer Depletion," Hearings before the U.S. House of Representatives, Committee on Energy and Commerce, Subcommittee on Health and the Environment, 9 March 1987, pp. 106–109.

The case for ozone depletion in 1986 was much stronger than it had been a decade earlier. It had become more difficult for responsible observers to argue that the concept of ozone loss had no scientific basis or was merely "media hype." Still, a number of authorities, especially those associated with industries that used CFCs and other ozone-depleting chemicals, continued to ask for moderation. They felt that actions to ban or limit the production of such chemicals was still premature. The key phrase that appears in much of their testimony before Congress and their statements to the general public was that "no imminent hazard" exists to human health or the environment. The following statements before the House Subcommittee on Health and the Environment are examples of this position.

The Position of the Fluorocarbon Program Panel of the Chemical Manufacturers Association, 1987

The Antarctic ozone phenomenon is a major, immediate concern. Research programs are in place which can, within a reasonable timescale, identify the mechanism(s) and evaluate the likelihood of this being a localized or more widespread phenomenon.

Timely support should be given to the deployment of the NDSC which is essential for the measurement and understanding of change in the stratosphere. A long term funding commitment by the international community is required.

Continued research and a better understanding of the science are needed to reduce uncertainties in predictions of future changes in atmospheric ozone and climate.

Current model calculations, incorporating the best available scientific information, indicate there is no imminent hazard from continued emissions of CFCs at or near today's rates. Thus, there is no scientific justification for a near-term phase-out or reduction in CFC emissions below today's production levels.

Source: "Ozone Layer Depletion," Hearings before the U.S. House of Representatives, Committee on Energy and Commerce, Subcommittee on Health and the Environment, 9 March 1987, p. 359.

Excerpt of Letter from Dr. Mary Good, Allied-Signal, Inc., 1987

- The Antarctic ozone phenomenon is of current concern. Research programs are in place which have the potential, within a reasonable time scale, to identify the mechanism(s) and evaluate the likelihood of this being a localized or more widespread phenomenon.
- Statistical analysis of ozone measurements from a ground-based network of stations over the past several decades shows no persistent change in total global amounts. It is this total amount of ozone overhead which determines the level of ultraviolet (UV) radiation reaching the earth's surface.
- Model calculations incorporating the best available scientific information indicate that there is no immediate hazard from continued emissions of CFC's at or about today's rates. Thus, we believe there is no scientific justification for a near-term phase-out or reduction in CFC emissions below today's production levels.
- But to be prudent, Allied-Signal supports *international* limits on the rate of growth of CFC production and/or emissions. Specifically, we support the efforts of the U.S. Environmental Protection Agency and the U.S. State Department working with the United Nations Environment Programme to achieve an international solution.
- However, unilateral U.S. action to control domestic production would provide only limited environmental protection, while making U.S. industry less competitive. It could even impede international resolution of what is a global problem. Even if CFC imports were restricted, foreign producers might see unilateral

U.S. regulation as an opportunity to increase their own market share in the rest of the world. Seizing this opportunity, they might resist international control and increase production, thereby defeating our basic goal.

- For our part, Allied-Signal is examining methods of reducing emissions in certain CFC applications. We are continuing to fund atmospheric research on this complex subject through the Fluorocarbon Program Panel of the Chemical Manufacturers Association. And we are looking for alternative CFC compounds that combine safety of use with acceptable environmental properties.

Source: "Ozone Layer Depletion," Hearings before the U.S. House of Representatives, Committee on Energy and Commerce, Subcommittee on Health and the Environment, 9 March 1987, pp. 229–230.

By the 1990s, articles disputing the loss of stratospheric ozone and the connection of that phenomenon with CFCs were being written primarily by observers with little or no background in the atmospheric sciences. In many cases, these observers were suggesting that the "ozone scare" was manufactured by individuals and organizations that had political agendas that could be advanced by promoting the notion of ozone depletion. In fact, it could be argued that such critics of the ozone depletion theory may have had their own political agenda in speaking out against the possibility of ozone loss.

The reader should be cautioned that the scientific arguments presented in the following excerpts may be subject to criticism by experts in the field of atmospheric sciences. For a more detailed analysis of the scientific errors presented in the works of Dixy Lee Ray, S. Fred Singer, Rush Limbaugh, and other critics writing in general books and periodicals, see the sidebar in *Chemical & Engineering News,* May 24, 1993, pp. 14–15.

Dixy Lee Ray in *Environmental Overkill*

Ray begins with a quotation from the 17 February 1992 cover story in *Time* magazine about ozone depletion. She argues that *Time*'s analysis of the issue is

> Nonsense. . . .Typical of the sort of "scientific" information available to the public, this article is so full of emotional hype, exaggeration, half-truths, and unsupported dogma that it is more propaganda than reporting. The only evidence that is referenced comes from the press releases of the National Aeronautics and

> Space Administration. And NASA, despite all our sentimentality about it, is hardly a reliable source. Like any other government agency, it is interested primarily only in those facts that justify more research and more federal dollars. The only scientists quoted in the *Time* magazine story are those who support the ozone-depletion-by-chlorofluorocarbons (CFCs) theory. No dissenting opinions, although there are plenty of them among reputable scientists, are mentioned. . . .

[Ray later attacks the notion that ozone depletion will be harmful to human health and to the environment.]

> The extent of ozone loss that EPA Director Reilly called "grim" would result in a 4 to 5 percent increase in exposure to ultraviolet radiation. This is, in fact, far less than normal, annual variations in ultraviolet radiation, and far less than the increased dosage one would absorb if one made a simple move to a lower latitude. . . .
>
> Who would refuse vacation in Hawaii, or Florida, or to the South Seas, or along the French Riviera because of such increases in the ultraviolet exposure? Yet these trips mean exposures many times in excess of the amounts that the EPA and NASA are using to frighten Americans.

Source: Dixy Lee Ray, *Environmental Overkill: Whatever Happened to Common Sense?* (Washington, DC: Regnery Gateway, 1993), pp. 28–29, 41–42.

From an Article by James P. Hogan, Novelist

> Periodically, societies are seized by collective delusions that take on lives of their own, where all facts are swept aside that fail to conform to the expectations of what has become a self-sustaining reality. Today we have the environmentalist mania reaching a crescendo over ozone.
>
> Manmade chlorofluorocarbons, or CFCs, we're told, are eating away the ozone layer that shields us from ultraviolet radiation, and if we don't stop using them now, deaths from skin cancer in the United States alone will rise by hundreds of thousands in the next half century. . . .
>
> Now, I'd have to agree that the alternative of seeing the planet seared by lethal levels of radiation [to finding substitutes for CFCs] would make a pretty good justification for whatever drastic action is necessary to prevent it. The only problem is, there isn't one piece of solid scientifically validated evidence to support the contention. The decisions being made are political, driven by media-friendly pressure groups wielding a power over public perceptions that is totally out of proportion to any scientific

> competence they possess. But when you ask the people who do have the competence to know — scientists who have specialized in the study of atmosphere and climate for years — a very different story emerges.

Source: James P. Hogan, "Ozone Politics: They Call This Science?" *Omni,* June 1993, pp. 35–38.

Selection from *The Hole in the Ozone Scare*

The book that appears to have been the single most influential document in the 1990s among critics of the CFC–ozone hole hypothesis is *The Hole in the Ozone Scare,* by Rogelio A. Maduro and Ralf Schauerhammer. The essence of the arguments posed in this book is summarized in a foreword to the book by the French volcanologist Haroun Tazieff. Tazieff writes that he became interested in the ozone depletion question and was "stimulated to get to the bottom of the question." What he discovered, he writes, was that:

> this [ozone] hole was most probably in existence at the time of the Shackleton expedition in 1909, and that most probably, it is a natural phenomenon. Consequently, it has nothing to do with CFCs, which did not exist at that time. . . .
>
> On the other hand the alarmist hypothesis, according to which CFCs are responsible for the Antarctic "hole," is based upon at least two acts of scientific bad faith, and that is sufficient to discredit it completely.
>
> The first is the deliberate omission of all reference to previous publications bearing on the same subject. . . .
>
> The second infringement of scientific ethics was that in their powerful and sustained effort (crowned by success) to make an impression upon the good public . . . the propagandists showed us satellite photographs of the Antarctic and of its "ozone hole" in the month of October only. . . . They did so without breathing a word to us about the fact that . . . [f]rom December on, this "hole" that the alarmists are using to sow panic, no longer exists.

Source: Rogelio A. Maduro and Ralf Schauerhammer, *The Hole in the Ozone Scare* (Washington: 21st Century Science Associates, 1992), p. xi.

"Who Lost the Ozone?"

For all the press given to contrarians like Ray, Hogan, and Maduro and Schauerhammer, concern about ozone depletion is still widespread among the vast majority of scientists who specialize in the atmospheric sciences. Journalists in the popular press also continue to draw the attention of the public to this issue which, many argued, was certainly not completely solved by the Montreal Protocol and later amendments to that pact. Typical of such arguments is the one summarized below. In an article somewhat similar to that from 1975 in *Science Digest*, quoted at the beginning of this section, writer Eugene Linden asks in a 1993 issue of *Time* magazine, "Who Lost the Ozone?" After a review of efforts made to cut back on CFC production over the previous six years, Linden argues that the only problem with these efforts is that:

> the rescuers may have arrived too late. No matter how quickly manufacturers halt the production of CFCs, billions of pounds of the chemical already produced will continue to seep into the atmosphere and rise inexorably to attack the ozone layer. Worse, measurement after measurement since the mid-1980s has shown that ozone loss has been greater and more rapid than scientists predicted. . . .

[Linden concludes with a fundamental observation about global problems such as ozone depletion:]

> The ozone story shows what can happen when the world underestimates problems. It also underscores the difficulty of imposing environmental regulations that clash with economic interests, especially in the face of scientific uncertainty. If policymakers wait until there is unarguable evidence of danger before they act, it may be too late to prevent serious environmental damage.

Source: Eugene Linden, "Who Lost the Ozone?" *Time*, 10 May 1993, pp. 56–58.

Documents

World Plan of Action

One of the earliest international statements about the possibility of ozone depletion is the World Plan of Action, adopted at a meeting of experts on ozone science in Washington, DC, in March of 1977. The meeting was sponsored by the United Nations Environment Programme. The following statement outlines the view of these experts about ozone depletion at the end of the 1970s. A key to organization acronyms appears at the end of this section.

The natural stratosphere and its ozone layer have been the subject of extensive research over the last fifty years. The need for jet aircraft to fly at high altitudes has long spurred the interest of the world's meteorological community. The curiosity of atmospheric physicists and chemists has provided the main thrust to illustrate the dynamics and photochemistry of the ozone layer.

In recent years research, associated with man's accelerating activities in the stratosphere and aerospace, has greatly modified our perception of ozone photochemistry. It is widely accepted that natural ozone photochemistry is controlled to a large extent by the action of nitric oxide derived from nitrous oxide which emanates from the earth's surface as a result of denitrification processes. This has added a new dimension to the ozone problem and new interest in studies of the earth's nitrogen cycle. It is now thought that stratospheric ozone is influenced by a variety of naturally occurring substances such as methane and methyl chloride. A continuing long-term programme is essential to our further understanding of these natural phenomena and as a basis for evaluating new effects.

Quite recently the impact of other substances such as the nitrogen oxides from aircraft exhaust and the man-made chlorofluoromethanes (CFMs) has been receiving intense scrutiny. As these are all trace substances with atmospheric concentrations as low as a fraction of a part per billion, the problems of measurement analysis and modelling have changed by an order of magnitude from those encountered when our understanding was based on the assumption of the simple Chapman (pure oxygen) system. Our knowledge and understanding have been greatly increased by a number of intensive national programmes, and there is a large measure of agreement on the model predictions that current aircraft emissions have minimal effects on the ozone layer but that the effects of continuing emissions of CFMs at the 1973 or higher levels are a matter of concern.

There are many gaps in our knowledge of factors affecting the ozone layer, and there may be factors that are as yet unrecognized. An intensive and well coordinated monitoring and research programme related to the occurrence of trace substances in the atmosphere, to test the model predictions and narrow their range of uncertainty, is particularly important.

Recommendations

The coordinated research and monitoring programme already initiated by the World Meteorological Organization, to clarify the basic dynamical photochemical and radiative aspects of the ozone layer and to evaluate the impact of man's activities on the ozone balance should be encouraged and supported.

Specifically this should include action to:

[The plan then describes in some detail the kinds of activities that should be carried out to learn more about ozone depletion. The headings that are listed are followed by brief descriptions of the kinds of activities to be followed in each case. Only the first section is reprinted in its entirety here. The acronym WMO indicates that that agency is to have primary responsibility for each activity.]

Monitor Ozone (WMO). Design, develop and operate an improved system (including appropriate ground-based, airborne and satellite subsystems) for monitoring and prompt reporting of global ozone, its vertical, spatial and temporal variations, with sufficient accuracy to detect small but statistically significant long-term trends.

Monitor solar radiation (WMO). . . .

Simultaneous species measurements (WMO). . . .

Chemical reactions (WMO). . . .

Development of computational modeling (WMO). . . .

Large-scale atmospheric transport (WMO). . . .

Global constituent budgets (WMO). . . .

The Impact of Changes in the Ozone Layer on Man and the Biosphere

A prediction of depletion of ozone due to man's activities would have little meaning unless it could be shown, at least qualitatively, that it would have significant effects on man and his environment.

It has been demonstrated that excessive exposure to ultraviolet radiation at 254 nm (radiation of a short [*sic*] wavelength than the UV-B which reaches the ground) causes tumors in laboratory animals and has deleterious effects on certain plants. Extensions of the studies to the UV-B band (from 290–320 nm) have been made by theoretical and experimental work and by a series of epidemiological studies. There is some evidence that increased UV-B would be associated with an increase in skin cancer and possibly in eye damage in susceptible sections of the human population. It is also likely that large increases in UV-B would damage nucleic acids and proteins and thereby have deleterious effects on plant and aquatic communities, but the effect of smaller changes in UV-B is highly uncertain. Although it is clear that significant progress has been made in a wide range of related research activities in the biological and medical sciences there are still many gaps in our knowledge of the effects of increased UV-B.

Recommendations

A wide variety of investigations of the impact of ozone depletion and increased ultraviolet radiation (UV-B) on man and the biosphere should be encouraged, supported and coordinated. Specifically it is proposed that action be undertaken on:

UV radiation. (i) *Monitor UV-B radiation* (WMO/WHO, FAO) . . . (ii) *Develop UV-B instrumentation* (WMO/UNEP, WHO, FAO) . . .(iii) *Promote UV-B research* (WMO, WHO, FAO). . . .

Human health. (i) *Statistics on skin cancer* (WHO) . . . (ii) *Research on induction mechanisms* (WHO) . . . (iii) *Other health aspects* (WHO). . . .

Other biological effects. (i) *Responses to UV-B* (FAO) . . . (ii) *Plant communities* (FAO) . . . (iii) *Aquatic systems* (FAO). . . .

Effects on climate. (i) *development* [sic] *of computational modelling* (WMO) . . . (ii)*Regional climate* (WMO/FAO). . . .

Socio-Economic Aspects

To provide a useful contribution to policy formulation, the assessment of a potential environmental hazard must include an evaluation of the costs and benefits to society that would result from a reduction of the hazard. Neither the methodology nor the available data are adequate to properly assess the socio-economic effects of stratospheric pollution or of measures taken to control it. Much work has been done on the costs of alternative courses of action particularly with respect of possible limitations to total CFM emissions. However, further elaboration is needed before a complete evaluation can be made with confidence.

Recommendations

Studies of the socio-economic impact of predicted ozone layer depletions and of alternative courses of actions to limit or control identified ozone-depleting emissions to the atmosphere should be supported at the national and international level. Specifically it is proposed that action be undertaken to [develop]:

Production and emission data (UNEP, ICC, OECD, ICAO). . . .

Methodology (UNESA, OECD). . . .

Institutional Arrangements

1. The Action Plan will be implemented by UN bodies, specialized agencies, international, national, intergovernmental, and non-governmental organizations and scientific institutions.
2. The UNEP should exercise a broad coordinating and catalytic role aimed at the integration and coordination of research efforts by arranging for:
   —collation of information on ongoing and planned research activities;
   —presentation and review of the results of research;
   —identification of further research needs.
3. In order for UNEP to fulfill that responsibility it should establish a Coordinating Committee on Ozone Matters composed of representatives of the agencies and non-governmental organizations participating in implementing the Action Plan as well as representatives of countries which have major scientific programmes contributing to the Action Plan. . . .
4. [This section deals with UNEP's responsibilities for coordinating national activities.]
5. [This section deals with staffing needed to carry out the Plan.]
6. UNEP should consider the need for and feasibility of establishing special coordinating mechanisms or

procedures for certain areas of interdisciplinary work, such as photobiology, which presently lack such coordinating facilities.

[Key to acronyms:]
FAO: Food and Agricultural Organization
ICAO: International Civil Aviation Organization
ICC: International Chamber of Commerce
OECD: Organization for Economic Cooperation and Development
UNEP: United Nations Environment Programme
UNESA: United Nations Department of Economic and Social Affairs
WHO: World Health Organization
WMO: World Meteorological Organization

Source: Asit K. Biswas, ed. *The Ozone Layer* (Oxford: Pergamon Press, 1979), pp. 377–381.

Vienna Convention

The first major international agreement about protection of the ozone layer was the Vienna Convention for the Protection of the Ozone Layer, signed in Vienna in March 1985. Of the 43 nations represented at Vienna, 20 signed at the convention's close on 22 March. As of 31 July 1993, an additional 105 nations had ratified the convention. Excerpts from that document are reprinted below.

Preamble

The Parties to this convention,

Aware of the potentially harmful impact on human health and the environment through modification of the ozone layer,

Recalling the pertinent provisions of the Declaration of the United Nations Conference on the Human Environment, and in particular principle 21, which provides that "States have, in accordance with the Charter of the United Nations and the principles of international law, the sovereign right to exploit their own resources pursuant to their own environmental policies, and the responsibility to ensure that activities within their jurisdiction or control do not cause damage to the environment of other States or of areas beyond the limits of national jurisdiction,"

Taking into account the circumstances and particular requirements of developing countries,

Mindful of the work and studies proceeding within both international and national organizations and, in particular, of the World Plan of Action on the Ozone Layer of the United Nations Environment Programme,

Mindful also of the precautionary measures for the protection of the ozone layer which have already been taken at the national and international levels,

Aware that measures to protect the ozone layer from modifications due to human activities require international co-operation and action, and should be based on relevant scientific and technical considerations,

Aware also of the need for further research and systematic observations to further develop scientific knowledge of the ozone layer and possible adverse effects resulting from its modification,

Determined to protect human health and the environment against adverse effects resulting from modification of the ozone layer,

HAVE AGREED AS FOLLOWS:

Article 1. Definitions

[This section defines seven terms used in the agreement, terms such as "ozone layer," "adverse effects," and "alternative technologies or equipment."]

Article 2. General Obligations

1. The Parties shall take appropriate measures in accordance with the provisions of this Convention and of those protocols in force to which they are party to protect human health and the environment against adverse effects resulting or likely to result from human activities which modify or are likely to modify the ozone layer.
2. To this end the Parties shall, in accordance with the means at their disposal and their capabilities:

(a) Co-operate by means of systematic observations, research and information exchange in order to better understand and assess the effects of human activities on the ozone layer and the effects on human health and the environment from modification of the ozone layer;
(b) Adopt appropriate legislative or administrative measures and co-operate in harmonizing appropriate policies to control, limit, reduce or prevent human activities under their jurisdiction or control should it be found that these activities have or are likely to have adverse effects resulting from modification or likely modification of the ozone layer;
(c) Co-operate in the formulation of agreed measures, procedures, and standards for the implementation of this convention, with a view to the adoption of protocols and annexes;
(d) Co-operate with competent international bodies to implement effectively this Convention and protocols to which they are party.

3. The provisions of this Convention shall in no way affect the right of Parties to adopt, in accordance with international law, domestic measures additional to those referred to in paragraphs 1 and 2 above, nor shall they affect additional domestic measures already taken by a Party, provided that these measures are not incompatible with their obligations under this convention.
4. The application of this article shall be based on relevant scientific and technical considerations.

Article 3. Research and Systematic Observations

1. The Parties undertake, as appropriate, to initiate and co-operate in, directly or through competent international bodies, the conduct of research and scientific assessments on:
   (a) The physical and chemical processes that may affect the ozone layer;
   (b) The human health and other biological effects deriving from any modifications of the ozone layer, particularly those resulting from changes in ultra-violet solar radiation having biological effects (UV-B);
   (c) Climatic effects deriving from any modifications of the ozone layer;

(d) Effects deriving from any modifications of the ozone layer and any consequent changes in UV-B radiation on natural and synthetic material useful to mankind;
(e) Substances, practices, processes and activities that may affect the ozone layer, and their cumulative effects;
(f) Alternative substances and technologies;
(g) Related socio-economic matters;

and as further elaborated in annexes I and II.

2. The Parties undertake to promote or establish, as appropriate, directly or through competent international bodies and taking fully into account national legislation and relevant ongoing activities at both the national and international levels, joint or complementary programmes for systematic observation of the state of the ozone layer and other relevant parameters, as elaborated in annex I.
3. The Parties undertake to co-operate, directly or through competent international bodies, in ensuring the collection,validation and transmission of research and observational data through appropriate world data centres in a regular and timely fashion.

Article 4. Co-operation in the Legal, Scientific and Technical Fields

[This section describes some of the ways in which the Parties to the agreement will be expected to work together in dealing with ozone layer–related issues. Included among these ways are the following:]

(a) Facilitation of the acquisition of alternative technologies by other Parties;
(b) Provision of information on alternative technologies and equipment, and supply of special manuals or guides to them;
(c) The supply of necessary equipment and facilities for research and systematic observations;
(d) Appropriate training of scientific and technical personnel.

Article 5. Transmission of Information

The Parties shall transmit, through the secretariat, to the Conference of the Parties established under article 6 information on the measures adopted by them in implementation of this Convention and of protocols to which they are party in such form and at such intervals as the meetings of the parties to the relevant instruments may determine.

Article 6. Conferences of the Parties

[This section outlines many of the mechanics by which the Vienna Convention is to carried out. Some pertinent provisions include the following:]

1. A Conference of the Parties is hereby established. . . .

. . .

4. The Conference of the Parties shall keep under continuous review the implementation of this Convention, and, in addition, shall: . . .

   . . .

   (b) Review the scientific information on the ozone layer, on its possible modification and on possible effects of any such modification;
   (c) Promote, in accordance with article 2, the harmonization of appropriate policies, strategies and measures for minimizing the release of substances causing or likely to cause modification of the ozone layer, and make recommendations on any other measures relating to this Convention;
   (d) adopt, in accordance with articles 3 and 4, programmes for research, systematic observations, scientific and technological co-operation, the exchange of information and the transfer of technology and knowledge;

   . . .

   (i) Establish such subsidiary bodies as are deemed necessary for the implementation of this Convention;
   (j) Seek, where appropriate, the services of competent international bodies and scientific committees, in particular the World Meteorological Organization and the World Health Organization, as well as the Co-ordinating Committee on the Ozone Layer, in scientific research, systematic observations and other activities pertinent to the objectives of this convention, and make use as appropriate of information from these bodies and committees; . . .

Article 7. Secretariat

[This section establishes the administrative body under which the Convention is to be carried out.]

Article 8. Adoption of Protocols

[This section allows the Conference of the Parties to adopt protocols that will carry out the objectives of the Convention.]

Article 9. Amendment of the Convention or Protocols

[This section describes the manner in which amendments to the Convention may be made. The key section is the following.]

3. The Parties shall make every effort to reach agreement on any proposed amendment to this Convention by consensus. If all efforts at consensus have been exhausted, and no agreement reached, the amendment shall as a last resort be adopted by a three-fourths majority of the Parties present and voting at the meeting, and shall be submitted by the Depositary to all Parties for ratification, approval or acceptance.

Article 10. Adoption and Amendment of Annexes

[This section describes the procedures by which annexes to the Convention may be adopted or amended.]

Article 11. Settlement of Disputes

[This section provides for the settlement of disputes among the Parties by, first, negotiation and then, if negotiation fails, by arbitration.]

Article 12. Signature

This Convention shall be open for signature at the Federal Ministry for Foreign Affairs of the Republic of Austria in Vienna from 22 March 1985 to 21 September 1985, and at United

Nations Headquarters in New York from 22 September 1985 to 21 March 1986.

Article 13. Ratification, Acceptance or Approval

1. This Convention and any protocol shall be subject to ratification, acceptance or approval by States and by regional economic integration organizations. Instruments of ratification, acceptance or approval shall be deposited with the Depositary. . . .

Article 14. Accession

1. This Convention and any protocol shall be open for accession by States and by regional economic integration organizations from the date on which the convention or the protocol concerned is closed for signature. The instruments of accession shall be deposited with the Depositary. . . .

Article 15. Right to Vote

1. Each Party to this Convention or to any protocol shall have one vote.

[Some details about voting rights follow.]

Article 16. Relationship between the Convention and Its Protocols

1. A State or a regional economic integration organization may not become a part to a protocol unless it is, or becomes at the same time, a Party of the Convention.
2. Decisions concerning any protocol shall be taken only by the parties to the protocol concerned.

Article 17. Entry into Force

1. This convention shall enter into force on the ninetieth day after the date of deposit of the twentieth instrument of ratification, acceptance, approval or accession.
2. Any protocol, except as otherwise provided in such protocol, shall enter into force on the ninetieth day after the

date of deposit of the eleventh instrument of ratification, acceptance or approval of such protocol or accession thereto. . . .

Article 18. Reservations

No reservations may be made to this Convention.

Article 19. Withdrawal

1. At any time after four years from the date on which this Convention has entered into force for a Party, that Party may withdraw from the Convention by giving written notification to the Depositary.

[More details about the process of withdrawal follow.]

Article 20. Depositary

1. The Secretary-General of the United Nations shall assume the functions of Depositary of this Convention and any protocols.

[The duties of the Depositary are explained in the following sections.]

Article 21. Authentic Texts

The original of this Convention, of which the Arabic, Chinese, English, French, Russian and Spanish texts are equally authentic, shall be deposited with the Secretary-General of the United Nations.

IN WITNESS WHEREOF the undersigned, being duly authorized to that effect, have signed this Convention.

DONE at Vienna on the 22nd day of March 1985.

There are two annexes to the treaty. Annex I, "Research and Systematic Observations," outlines the nature of the research on the ozone layer that is to be carried out by parties to the Vienna Convention. Some examples include studies of the physics and chemistry of the atmosphere, health and biological effects of ozone depletion, and effects of ozone depletion on climate. Annex II describes the kinds of information that will be exchanged among signatories and the mechanisms by which those exchanges will take place.

Montreal Protocol

The Vienna Convention, cited above, was only one step on the way to the most important international agreement of all, the Montreal Protocol on Substances That Deplete the Ozone Layer. That agreement was signed in Montreal on 16 September 1987 by 25 nations. The protocol was eventually ratified by 122 nations. As a result of regular meetings held by the signatories to the protocol, the original document has changed many times. These changes have reflected increasing scientific information about the ozone layer and changes in the production and consumption of CFCs and other ozone-depleting chemicals. Portions of the original protocol are quoted below. While the most recent version of the protocol looks quite different from the original version, the fundamental concepts adopted in 1987 are still retained in updated versions.

Preamble

The Parties to this Protocol,

Being Parties to the Vienna Convention for the Protection of the Ozone Layer,

Mindful of their obligation under that Convention to take appropriate measures to protect human health and the environment against adverse effects resulting or likely to result from human activities which modify or are likely to modify the ozone layer,

Recognizing that world-wide emissions of certain substances can significantly deplete and otherwise modify the ozone layer in a manner that is likely to result in adverse effects on human health and the environment,

Conscious of the potential climatic effects of emissions of these substances,

Aware that measures taken to protect the ozone layer from depletion should be based on relevant scientific knowledge, taking into account technical and economic considerations,

Determined to protect the ozone layer by taking precautionary measures to control equitably total global emissions of substances that deplete it, with the ultimate objective of their elimination on the basis of developments in scientific knowledge, taking into account technical and economic considerations,

Acknowledging that special provision is required to meet the needs of developing countries for these substances,

Noting the precautionary measures for controlling emissions of certain chlorofluorocarbons that have already been taken at national and regional levels,

Considering the importance of promoting international co-operation in the research and development of science and technology relating to the control and reduction of emissions of substances that deplete the ozone layer, bearing in mind in particular the needs of developing countries,

HAVE AGREED AS FOLLOWS:

Article 1. Definitions

[The first section defines eight terms of major significance used in the protocol. They are: convention, parties, secretariat, controlled substances, production, consumption, calculated levels, and industrial rationalization.]

Article 2. Control Measures

[This section is key to carrying out the intentions of the signers of the protocol. Its most important features are extracted below.]

1. Each Party shall ensure that for the twelve-month period commencing on the first day of the seventh month following the date of the entry into force of this Protocol, and in each twelve-month period thereafter, its calculated

level of consumption of the controlled substances in Group I of Annex A does not exceed its calculated level of consumption in 1986. By the end of the same period, each Party producing one or more of these substances shall ensure that its calculated level of production of the substances does not exceed its calculated level of production in 1986, except that such level may have increased by no more than ten per cent based on the 1986 level. Such increase shall be permitted only so as to satisfy the basic domestic needs of the Parties operating under Article 5 and for the purposes of industrial rationalization between Parties.

2. [This paragraph repeats the above restrictions for controlled substances in Group II of Annex A.]
3. Each Party shall ensure that for the period 1 July 1993 to 30 June 1994 and in each twelve-month period thereafter, its calculated level of consumption of the controlled substances in Group I of Annex A does not exceed annually eighty per cent of its calculated level of consumption in 1986. [The same restriction is then applied to the production of the controlled substances in Group II of Annex A.]
4. Each Party shall ensure that for the period 1 July 1998 to 30 June 1999, and in each twelve-month period thereafter, its calculated level of consumption of the controlled substances in Group I of Annex A does not exceed annually, fifty per cent of its calculated level of consumption in 1986. [The same restriction is then applied to the production of the controlled substances in Group I of Annex A.] This paragraph will apply unless the Parties decide otherwise at a meeting by a two-thirds majority of Parties present and voting, representing at least two-thirds of the total calculated level of consumption of these substances of the Parties. This decision shall be considered and made in the light of the assessments referred to in Article 6.
5. Any Party whose calculated level of production in 1986 of the controlled substances in Group I of Annex A was less than twenty-five kilotonnes may, for the purposes of industrial rationalization, transfer to or receive from any other Party, production in excess of the limits set out in paragraphs 1, 3 and 4 provided that the total combined calculated level of production of the Parties concerned does not exceed the production limits set out in this

Article. Any transfer of such production shall be notified to the secretariat, no later than the time of the transfer.

6. Any Party not operating under Article 5, that has facilities for the production of controlled substances under construction, or contracted for, prior to 16 September 1987, and provided for in national legislation prior to 1 January 1987, may add the production from such facilities to the 1986 production of such substances for the purposes of determining its calculated level of production for 1986, provided that such facilities are completed by 31 December 1990 and that such production does not raise that Party's annual calculated level of consumption of the controlled substances above 0.5 kilograms per capita.
7. [This section provides for notification of the secretariat in cases described in section 6.]
8. [This section refers to the ways in which regional economic integration organizations, such as the European Economic Community, are to deal with provisions of the preceding sections.]
9. [This section describes the manner in which adjustments are to be made, pursuant to Article 6 of the protocol, by parties to the agreement. The decision-making process is described in sub-section (c), as follows:]
   (c) In taking such decision, the Parties shall make every effort to reach agreement by consensus. If all efforts at consensus have been exhausted, and no agreement reached, such decisions shall, as a last resort, be adopted by a two-thirds majority of the Parties present and voting representing at least fifty per cent of the total consumption of the controlled substances of the Parties.
10. (a) Based on the assessments made pursuant to Article 6 of this Protocol and in accordance with the procedure set out in Article 9 of the Convention, the Parties may decide:
      (i) whether any substances, and if so which, should be added to or removed from any annex to this Protocol; and
      (ii) the mechanism, scope and timing of the control measures that should apply to those substances;
   (b) Any such decision shall become effective, provided that it has been accepted by a two-thirds majority vote of the Parties present and voting.

11. Notwithstanding the provisions contained in this Article, Parties may take more stringent measures than those required by this Article.

Article 3. Calculation of Control Levels

For the purposes of Articles 2 and 5, each Party shall, for each Group of substances in Annex A, determine its calculated levels of:

(a) production by:
  (i) multiplying its annual production of each controlled substance by the ozone depleting potential specified in respect of it in Annex A; and
  (ii) adding together, for each such Group, the resulting figures;

(b) imports and exports, respectively, by following, *mutatis mutandis*, the procedure set out in subparagraph (a); and

(c) consumption by adding together its calculated levels of production and imports and subtracting its calculated level of exports as determined in accordance with subparagraphs (a) and (b). However, beginning on 1 January 1993, any export of controlled substances to non-Parties shall not be subtracted in calculating the consumption level of the exporting Party.

Article 4. Control of Trade with Non-Parties

[This section outlines the kinds of trades of ozone-depleting materials that are and are not permitted by the protocol. For example:]

1. Within one year of the entry into force of this Protocol, each Party shall ban the import of controlled substances from any State not party to this Protocol.

. . .

5. Each Party shall discourage the export, to any State not party to this Protocol, of technology for producing and for utilizing controlled substances.
6. Each Party shall refrain from providing new subsidies, aid, credits, guarantees or insurance programmes for the

export to States not party to this Protocol of products, equipment, plants or technology that would facilitate the production of controlled substances. . . .

Article 5. Special Situation of Developing Countries

[A delicate point of negotiation in debate over the protocol was how to handle the special needs of developing nations. The general principle that was developed is enunciated in this section.]

1. Any Party that is a developing country and whose annual calculated level of consumption of the controlled substances is less than 0.3 kilograms per capita on the date of the entry into force of the Protocol shall, in order to meet its basic domestic needs, be entitled to delay its compliance with the control measures set out in paragraphs 1 to 4 of Article 2 by ten years after that specified in these paragraphs. However, such Party shall not exceed an annual calculated level of consumption of 0.3 kilograms per capita. Any such Party shall be entitled to use either the average of its annual calculated level of consumption for the period 1995 to 1997 inclusive or a calculated level of consumption of 0.3 kilograms per capita, whichever is the lower, as the basis for its compliance with the control measures. . . .

Article 6. Assessment and Review of Control Measures

Beginning in 1990, and at least every four years thereafter, the Parties shall assess the control measures provided for in Article 2 on the basis of available scientific, environmental, technical and economic information. At least one year before each assessment, the Parties shall convene appropriate panels of experts qualified in the fields mentioned and determine the composition and terms of reference of any such panels. Within one year of being convened, the panels will report their conclusions, through the secretariat, to the Parties.

Article 7. Reporting of Data

[This section provides for the reporting of production and consumption data by Parties to the secretariat.]

Article 8. Non-Compliance

The Parties, at their first meeting, shall consider and approve procedures and institutional mechanisms for determining non-compliance with the provisions of this Protocol and for treatment of Parties found to be in non-compliance.

Article 9. Research, Development, Public Awareness and Exchange of Information

1. The Parties shall co-operate, consistent with their national laws, regulations and practices and taking into account in particular the needs of developing countries, in promoting, directly or through competent international bodies, research, development and exchange of information on:
   (a) best technologies for improving the containment, recovery, recycling or destruction of controlled substances or otherwise reducing their emissions;
   (b) possible alternatives to controlled substances, to products containing such substances, and to products manufactured with them; and
   (c) costs and benefits of relevant control strategies. . . .

Article 10. Technical Assistance

[This article declares that Parties to the protocol have a special obligation to provide technical assistance to each other, but especially to developing countries, for meeting the conditions set out in the protocol.]

Article 11. Meetings of the Parties

[This article provides that signers of the protocol are to meet regularly, the first meeting being not later than one year after the protocol takes effect. An extensive and detailed description is then provided of the rules and conditions for such meetings.]

Article 12. Secretariat

[This article defines the duties and responsibilities of the secretariat, the administrative body responsible for the ongoing operation of the protocol.]

Article 13. Financial Provisions

1. The funds required for the operation of this Protocol, including those for the functioning of the secretariat related to this Protocol, shall be charged exclusively against contributions from the Parties.
2. The Parties, at their first meeting, shall adopt by consensus financial rules for the operation of this Protocol.

Article 14. Relationships of This Protocol to the Convention

Except as otherwise provided in this Protocol, the provisions of the Convention relating to its protocols shall apply to this Protocol.

Article 15. Signature

This Protocol shall be open for signature by States and by regional economic integration organizations in Montreal on 16 September 1987, in Ottawa from 17 September 1987 to 16 January 1988 and at the United Nations Headquarters in New York from 17 January 1988 to 15 September 1988.

Article 16. Entry into Force

1. This Protocol shall enter into force on 1 January 1989, provided that at least eleven instruments of ratification, acceptance, approval of the Protocol or accession thereto have been deposited by States or regional economic integration organizations representing at least two-thirds of 1986 estimated global consumption of the controlled substances, and the provisions of paragraph 1 of Article 17 of the Convention have been fulfilled. In the event that these conditions have not been fulfilled by that date, the Protocol shall enter into force on the ninetieth day

following the date on which the conditions have been fulfilled. . . .

Article 17. Parties Joining after Entry into Force

Subject to Article 5, any State or regional economic integration organization which becomes a Party to this Protocol after the date of its entry into force, shall fulfil forthwith the sum of the obligations under Article 2, as well as under Article 4, that apply at that date to the States and regional economic integration organizations that became Parties on the date the Protocol entered into force.

Article 18. Reservations

No reservations may be made to this Protocol.

Article 19. Withdrawal

For the purposes of this Protocol, the provisions of Article 19 of the Convention relating to withdrawal shall apply, except with respect to Parties referred to in paragraph 1 of Article 5. Any such Party may withdraw from this Protocol by giving written notification to the Depositary at any time after four years of assuming the obligations specified in paragraphs 1 to 4 of Article 2. Any such withdrawal shall take effect upon expiry of one year after the date of its receipt by the Depositary, or on such later date as may be specified in the notification of the withdrawal.

Article 20. Authentic Texts

The original of this Protocol, of which the Arabic, Chinese, English, French, Russian and Spanish texts are equally authentic, shall be deposited with the Secretary-General of the United Nations.

IN WITNESS WHEREOF the undersigned, being duly authorized to that effect, have signed this protocol.

DONE at Montreal this sixteenth day of September, one thousand nine hundred and eighty-seven.

Annex A. Controlled Substances

Group	Substance	Ozone-depleting potential[a]
Group I		
$CFCl_3$	(CFC 11)	1.0
CF_2Cl_2	(CFC 12)	1.0
$C_2F_3Cl_3$	(CFC 113)	0.8
$C_2F_4Cl_2$	(CFC 114)	1.0
CF_5Cl	(CFC 115)	0.6
Group II		
CF_2BrCl	halon 1211	3.0
CF_3Br	halon 1301	10.0
$C_2F_4Br_2$	halon 2402	(to be determined)

[a]Ozone-depleting potentials are estimates based on existing knowledge and will be reviewed and revised periodically.
Source: Montreal Protocol on Substances That Deplete the Ozone Layer, Final Act (Nairobi, Kenya: United Nations Environment Programme, 1987).

London Amendments to the Montreal Protocol

The signers of the Montreal Protocol met again in London in June 1990, as provided in the protocol, to assess recent studies on ozone depletion and to decide what changes to make in the original protocol. The summary of that meeting, "Report of the Second Meeting of the Parties to the Montreal Protocol on Substances That Deplete the Ozone Layer" is a very long document. The key points of agreement reached at the London meeting are extracted below.

Annex I. Adjustments to the Montreal Protocol on Substances That Deplete the Ozone Layer

[After introductory matter explaining how the Montreal Protocol is referenced in this document, the text goes on to revisions in the original document.]

A. Article 2A: CFC's

[This article begins with an explanation of the renumbering of sections necessary in the original protocol. It then makes the following changes in those sections:]

2. Each Party shall ensure that for the period from 1 July 1991 to 31 December 1992 its calculated levels of consumption and production of the controlled substances in Group I of Annex A do not exceed 150 per cent of its calculated levels of production and consumption of those substances in 1986; . . .
3. Each Party shall ensure that for the twelve-month period commencing on 1 January 1995, and in each twelve-month period thereafter, its calculated level of consumption of the controlled substances in Group I of Annex A does not exceed, annually, fifty per cent of its calculated level of consumption in 1986. Each Party producing one or more of these substances shall, for the same periods, ensure that its calculated level of production of the substances does not exceed, annually, fifty per cent of its calculated level of production in 1986. However, in order to satisfy the basic domestic needs of the Parties operating under paragraph 1 of Article 5, its calculated level of production may exceed that limit by up to ten per cent of its calculated level of production in 1986.
4. Each Party shall ensure that for the twelve-month period commencing on 1 January 1997, and in each twelve-month period thereafter, its calculated level of consumption of the controlled substances in Group I of Annex A does not exceed, annually, fifteen per cent of its calculated level of consumption in 1986. Each Party producing one or more of these substances shall, for the same periods, ensure that its calculated level of production of the substances does not exceed, annually, fifteen per cent of its calculated level of production in 1986. However, in order to satisfy the basic domestic needs of the Parties operating under paragraph 1 of Article 5, its calculated level of production may exceed that limit by up to ten per cent of its calculated level of production in 1986.
5. Each Party shall ensure that for the twelve-month period commencing on 1 January 2000, and in each twelve-month period thereafter, its calculated level of consumption of the controlled substances in Group I of Annex A does not exceed zero. Each Party producing one or more of these substances shall, for the same periods, ensure that its calculated level of production of the substances does not exceed zero. However, in order to satisfy the basic domestic needs of the Parties operating under

paragraph 1 of Article 5, its calculated level of production may exceed that limit by up to fifteen per cent of its calculated level of production in 1986.

6. In 1992, the Parties will review the situation with the objective of accelerating the reduction schedule.

B. Article 2B: Halons

Paragraph 2 of Article 2 of the Protocol shall be replaced by the following paragraphs, which shall be numbered paragraphs 1 to 4 of Article 2B:

Article 2B: Halons

1. Each Party shall ensure that for the twelve-month period commencing on 1 January 1992, and in each twelve-month period thereafter, its calculated level of consumption of the controlled substances in Group II of Annex A does not exceed, annually, its calculated level of consumption in 1986. Each Party producing one or more of these substances shall, for the same periods, ensure that its calculated level of production of the substances does not exceed, annually, its calculated level of production in 1986. However, in order to satisfy the basic domestic needs of the Parties operating under paragraph 1 of Article 5, its calculated level of production may exceed that limit by up to ten per cent of its calculated level of production in 1986.
2. Each Party shall ensure that for the twelve-month period commencing on 1 January 1995, and in each twelve-month period thereafter, its calculated level of consumption of the controlled substances in Group II of Annex A does not exceed, annually, fifty per cent of its calculated level of consumption in 1986. Each Party producing one or more of these substances shall, for the same periods, ensure that its calculated level of production of the substances does not exceed, annually, fifty per cent of its calculated level of production in 1986. However, in order to satisfy the basic domestic needs of the Parties operating under paragraph 1 of Article 5, its calculated level of production may exceed that limit by up to ten per cent of its calculated level of production in 1986. This paragraph will apply save to the extent that the Parties decide to

permit the level of production or consumption that is necessary to satisfy essential uses for which no adequate alternatives are available.

3. Each Party shall ensure that for the twelve-month period commencing on 1 January 2000, and in each twelve-month period thereafter, its calculated level of consumption of the controlled substances in Group II of Annex A does not exceed zero. Each Party producing one or more of these substances shall, for the same periods, ensure that its calculated level of production of the substances does not exceed zero. However, in order to satisfy the basic domestic needs of the Parties operating under paragraph 1 of Article 5, its calculated level of production may exceed that limit by up to fifteen per cent of its calculated level of production in 1986. This paragraph will apply save to the extent that the Parties decide to permit the level of production or consumption that is necessary to satisfy essential uses for which no adequate alternatives are available.
4. By 1 January 1993, the Parties shall adopt a decision identifying essential uses, if any, for the purposes of paragraphs 2 and 3 of this Article. Such decision shall be reviewed by the Parties at their subsequent meetings.

Annex II. Amendment to the Montreal Protocol on Substances That Deplete the Ozone Layer

Article 1. Amendment

A. Preambular Paragraphs

1. The 6th preambular paragraph of the Protocol shall be replaced by the following:

Determined to protect the ozone layer by taking precautionary measures to control equitably total global emissions of substances that deplete it, with the ultimate objective of their elimination on the basis of developments in scientific knowledge, taking into account technical and economic considerations and bearing in mind the developmental needs of developing countries.

2. The 7th preambular paragraph of the Protocol shall be replaced by the following:

   Acknowledging that special provision is required to meet the needs of developing countries, including the provision of additional financial resources and access to relevant technologies, bearing in mind that the magnitude of funds necessary is predictable, and the funds can be expected to make a substantial difference in the world's ability to address the scientifically established problem of ozone depletion and its harmful effects.

3. The 9th preambular paragraph of the Protocol shall be replaced by the following:

   Considering the importance of promoting international co-operation in the research, development and transfer of alternative technologies relating to the control and reduction of emissions of substances that deplete the ozone layer, bearing in mind in particular the needs of developing countries.

B. Article 1. Definitions

[This article revises two of the definitions provided in the original protocol and adds one new definition.]

[Sections C through N deal with "housekeeping" clauses needed to bring the London agreement into full confirmation with the wording of the Montreal Protocol. They also introduce a group of new "controlled substances" to be covered by the protocol. These substances are listed as Annex B, Groups I, II, and III.]

O. Article 4. Control of Trade with Non-Parties

[This article describes in detail conditions involving trade in controlled substances between parties and non-parties to the protocol.]

P. Article 5. Special Situation of Developing Countries

[This article describes in even more detail than the original agreement the special problems faced by developing countries in trying to meet the letter and spirit of the Montreal Protocol and outlines some ways in which parties to the agreement can assist developing countries in meeting those conditions. See in particular section T below.]

[Sections Q, R, and S deal with modifications of the original protocol involving Article 6: Assessment and Review of Control Measures; Article 7: Reporting of Data; and Article 9: Research, Development, Public Awareness and Exchange of Information.]

T. Article 10. Financial Mechanism

[Another key feature of the London Revision was the establishment of a Multilateral Fund, designed to help developing countries meet the conditions of the Montreal Protocol. The basic features of this article are as follows:]

1. The Parties shall establish a mechanism for the purposes of providing financial and technical co-operation, including the transfer of technologies, to Parties operating under paragraph 1 of Article 5 of this Protocol to enable their compliance with the control measures set out in Articles 2A to 2E of the Protocol. . . .
2. The mechanism established under paragraph 1 shall include a Multilateral Fund. . . .
3. The Multilateral Fund shall:
   (a) Meet, on a grant or concessional basis as appropriate, and according to criteria to be decided upon by the Parties, the agreed incremental costs;
   (b) Finance clearing-house functions to:
      (i) Assist Parties operating under paragraph 1 of Article 5, through country specific studies and other technical co-operation, to identify their needs for co-operation;
      (ii) Facilitate technical co-operation to meet these identified needs;
      (iii) Distribute, as provided for in Article 9, information and relevant materials, and hold workshops, training sessions, and other related activities, for the benefit of Parties that are developing countries; and
      (iv) Facilitate and monitor other multilateral, regional and bilateral co-operation available to Parties that are developing countries;

(c) Finance the secretarial services of the Multilateral Fund and related support costs. . . .

6. The Multilateral Fund shall be financed by contributions from Parties not operating under paragraph 1 of Article 5 in convertible currency or, in certain circumstances, in kind and/or in national currency, on the basis of the United Nations scale of assessments. . . .
7. The Parties shall decide upon the programme budget of the Multilateral Fund for each fiscal period and upon the percentage of contributions of the individual Parties thereto.
8. Resources under the Multilateral Fund shall be disbursed with the concurrence of the beneficiary Party.
9. Decisions by the Parties under this Article shall be taken by consensus whenever possible. If all efforts at consensus have been exhausted and no agreement reached, decisions shall be adopted by a two-thirds majority vote of the Parties present and voting, representing a majority of the Parties operating under paragraph 1 of Article 5 present and voting and majority of the Parties not so operating present and voting.
10. The financial mechanism set out in this Article is without prejudice to any future arrangements that may be developed with respect to other environmental issues.

[Articles U through X deal with additional minor modifications of original sections of the protocol.]

Y. Annexes

The following annexes shall be added to the Protocol:

Annex B. Controlled Substances

Group	Substance	Ozone-depleting potential[a]
Group I		
CF_3Cl	CFC 13	1.0
C_2FCl_5	CFC 111	1.0
$C_2F_2Cl_4$	CFC 112	1.0
C_3FCl_7	CFC 211	1.0
$C_3F_2Cl_6$	CFC 212	1.0
$C_3F_3Cl_5$	CFC 213	1.0

Group	Substance	Ozone-depleting potential[a]
$C_3F_4Cl_4$	CFC 214	1.0
$C_3F_5Cl_3$	CFC 215	1.0
$C_3F_6Cl_2$	CFC 216	1.0
C_3F_7Cl	CFC 217	1.0
Group II		
CCl_4	carbon tetrachloride	1.1
Group III		
C_2H_3Cl3[a]	1,1,1-trichloroethane (methyl chloroform)	0.1

[a]This formula does not refer to 1,1,2-trichloroethane

Annex C. Transitional Substances

Group I	Substance
$CHFCl_2$	HCFC 21
CHF_2Cl	HCFC 22
CH_2FCl	HCFC 31
C_2HFCl_4	HCFC 121
$C_2HF_2Cl_3$	HCFC 122
$C_2HF_3Cl_2$	HCFC 123
C_2HF_4Cl	HCFC 124
$C_2H_2FCl_3$	HCFC 131
$C_2H_2F_2Cl_2$	HCFC 132
$C_2H_2F_3Cl$	HCFC 133
$C_2H_3FCl_2$	HCFC 141
$C_2H_3F_2Cl$	HCFC 142
C_2H_4FCl	HCFC 151
C_3HFCl_6	HCFC 221
$C_3HF_2Cl_5$	HCFC 222
$C_3HF_3Cl_4$	HCFC 223
$C_3HF_4Cl_3$	HCFC 224
$C_3HF_5Cl_2$	HCFC 225
C_3HF_6Cl	HCFC 226
$C_3H_2FCl_5$	HCFC 231
$C_3H_2F_2Cl_4$	HCFC 232
$C_3H_2F_3Cl_3$	HCFC 233
$C_3H_2F_4Cl_2$	HCFC 234
$C_3H_2F_5Cl$	HCFC 235
$C_3H_3FCl_4$	HCFC 241

Group I	Substance
$C_3H_3F_2Cl_3$	HCFC 242
$C_3H_3F_3Cl_2$	HCFC 243
$C_3H_3F_4Cl$	HCFC 244
$C_3H_4FCl_3$	HCFC 251
$C_3H_4F_2Cl_2$	HCFC 252
$C_3H_4F_3Cl$	HCFC 253
$C_3H_5FCl_2$	HCFC 261
$C_3H_5F_2Cl$	HCFC 262
C_3H_6FCl	HCFC 271

Article 2. Entry into Force

1. This Amendment shall enter into force on 1 January 1992, provided that at least twenty instruments of ratification, acceptance or approval of the Amendment have been deposited by States or regional Economic integration organizations that are Parties to the Montreal Protocol on Substances that Deplete the Ozone Layer. . . .

Additional Material

Annex IV. Appendix II. Terms of Reference of the Executive Committee

1. The Executive Committee of the Parties is established to develop and monitor the implementation of specific operational policies, guidelines and administrative arrangements including the disbursement of resources, for the purpose of achieving the objectives of the Multilateral Fund under the Financial Mechanism.
2. The Executive Committee shall consist of seven Parties from the group of Parties operating under paragraph 1 of Article 5 of the Protocol and seven Parties from the group of Parties not so operating. Each group shall select its Executive Committee members. The members of the Executive Committee shall be formally endorsed by the Meeting of the Parties.

[The remaining sections of the appendix describe the organization, financing, meetings, and functions of the Executive Committee.]

Annex IV. Appendix IV. Terms of Reference for the Interim Multilateral Fund

A. Establishment

1. An interim Multilateral Fund of $160 million, which could be raised by up to $80 million during the three-year period when more countries become Parties to the Protocol, hereinafter referred to as "the Multilateral Fund," shall be established.

B. Roles of the Implementing Agencies

[The next five sections describe the functions of the Executive Committee, the United Nations Environment Programme, the United Nations Development Programme, the World Bank, and other agencies involved in operating the Multilateral Fund.]

[Sections C and D deal with budget and administration of the fund, respectively.]

Annex VII. Resolution by the Governments and the European Communities Represented at the Second Meeting of the Parties to the Montreal Protocol

The Governments and the European Communities represented at the Second Meeting of the Parties to the Montreal Protocol

Resolve:

I. Other Halons Not Listed in Annex A, Group II of the Montreal Protocol ("Other Halons")

1. To refrain from authorizing or to prohibit production and consumption of fully halogenated compounds containing one, two or three carbon atoms and at least one atom each of bromine and fluorine [footnote omitted], and not listed in Group II of Annex A of the Montreal Protocol (hereafter called "other halons"), which are of such a chemical nature or such a quantity that they would pose a threat to the ozone layer;

2. To refrain from using other halons except for those essential applications where other more environmentally suitable alternative substances or technologies are not yet available; and
3. To report to the Secretariat to the Protocol estimates of their annual production and consumption of such other halons;

II. Transitional Substances

1. To apply the following guidelines to facilitate the adoption of transitional substances with a low ozone-depleting potential, such as hydrochlorofluorocarbons (HCFCs), where necessary, and their timely substitution by non-ozone depleting and more environmentally suitable alternative substances or technologies:
   (a) Use of transitional substances should be limited to those applications where other more environmentally suitable alternative substances or technologies are not available;
   (b) Use of transitional substances should not be outside the areas of application currently met by the controlled and transitional substances, except in rare cases for the protection of human life or human health;
   (c) Transitional substances should be selected in a manner that minimizes ozone depletion, in addition to meeting other environmental, safety and economic considerations;
   (d) Emission control systems, recovery and recycling should, to the degree possible, be employed in order to minimize emissions to the atmosphere;
   (e) Transitional substances should, to the degree possible, be collected and prudently destroyed at the end of their final use;
2. To review regularly the use of transitional substances, their contribution to ozone depleting and global warming, and the availability of alternative products and application technologies, with a view to their replacement by non–ozone-depleting and more environmentally suitable alternatives and as the scientific evidence requires: at present, this should be no later than 2040 and, if possible, no later than 2020;

III. 1,1,1-Trichloroethane (Methyl Chloroform)

1. To phase out production and consumption of methyl chloroform as soon as possible;
2. To request the Technology Review Panel to investigate the earliest technically feasible dates for reductions and total phase-out; and
3. To request the Technology Review Panel to report their findings to the preparatory meeting of the Parties with a view to the consideration by the Meeting of the Parties, not later than 1992;

IV. More Stringent Measures

1. To express appreciation to those Parties that have already taken measures more stringent and broader in scope than those required by the Protocol;
2. To urge adoption, in accordance with the spirit of paragraph 11 of Article 2 of the Protocol, of such measures in order to protect the ozone layer.

Statement by Heads of Delegations of the Governments of Australia, Austria, Belgium, Canada, Denmark, Federal Republic of Germany, Finland, Liechtenstein, Netherlands, New Zealand, Norway, Sweden, Switzerland

The heads of delegations of the above governments represented at the Second Meeting of the Parties to the Montreal Protocol

Concerned about the recent scientific findings on severe depletion of the ozone layer of both Southern and Northern Hemispheres;

Mindful that all CFCs are also powerful greenhouse gases leading to global warming,

Convinced of the availability of more environmentally suitable alternative substances or technologies, and

Convinced of the need to further tighten control measures of CFCs beyond the Protocol adjustments agreed by the Parties to the Montreal Protocol,

DECLARE their firm determination to take all appropriate measures to phase out the production and consumption of all fully halogenated chlorofluorocarbons controlled by the Montreal Protocol, as adjusted and amended, as soon as possible, but not later than 1997.

Source: Ozone Secretariat, *Handbook for the Montreal Protocol on Substances That Deplete the Ozone Layer,* Third Edition, August 1993.

1990 Amendment to Clean Air Act

One of the key documents outlining the way in which the United States would react to the issue of ozone depletion was the Clean Air Act Amendments of 1990. These amendments clearly outlined the threat posed by depletion of stratospheric ozone and the actions that would be available to government agencies in their response to this problem. Reproduced below is a detailed summary of Title VI of the act, "Stratospheric Ozone Protection."

Listing

Within 60 days of enactment, EPA to publish two lists of ozone-depleting substances:

(1) Class I substances are the most potent ozone depleters, include chlorofluorocarbons (CFC)-11, 12, 113, 114, 115 (Group I); halons(Group II); all other fully halogenated CFCs (Group III); carbon tetrachloride (Group IV); and methyl chloroform (Group V).
2) Class II substances are the hydrochlorofluorocarbons (HCFCs).

At least every three years, EPA to add to the list other substances that meet specified criteria. Anyone may petition EPA to add a substance to one of the lists; EPA shall either add the substance to the list or publish a denial within 180 days. EPA may not remove any substance from the Class I list and may only remove a substance from the Class II list to add it to the Class I list.

Ozone Depletion and Global Warming Potential

Simultaneously with publication of the limits, EPA to assign each listed substance to an ozone depletion potential

(ODP), chlorine and bromine loading potential and atmospheric lifetime. One year after enactment, EPA to publish a global warming potential (GWP) for each substance.

Reporting Requirements

Quarterly reports (or as determined by Administrator) of production, imports and exports of Class I and Class II substances are required to be submitted to EPA. EPA to issue reporting regulations within 270 days.

EPA is to report on the domestic and worldwide production, use and consumption of Class I and Class II substances to Congress every 3 years. Every 6 years EPA must report on any environmental and economic effects of strtosperic [*sic*] ozone depletion.

Reduction Requirements

Class I Substances

Production and consumption (defined as product plus imports minus exports) of Class I substances to be capped according to following schedule. Percentages refer to maximum allowable production and consumption as a percentage of the quantity of the substance produced or consumed by a person in the baseline year.

Year	Carbon Tetrachloride	Methyl Chloroform	Other Class Substances
1991	100%	100%	85%
1992	90%	100%	80%
1993	80%	90%	75%
1994	70%	85%	65%
1995	15%	70%	50%
1996	15%	50%	40%
1997	15%	50%	15%
1998	15%	50%	15%
1999	15%	50%	15%
2000	—	20%	—
2001	—	20%	—

Baseline year for methyl chloroform and carbon tetrachloride is 1989. Baseline year for the other Class I substances is 1986. Administrator to choose a representative year for the baseline for Class II substances. Administrator authorized to grant limited exemptions from the phaseout schedule for specified purposes, so long as such exemptions are consistent with the United States' obligations under the Montreal Protocol on Substances That Deplete the Ozone Layer. EPA to issue regulations within 10 months of enactment.

Class II Substances

New uses of HCFCs banned January 1, 2015 unless the HCFCs are used, recovered and recycled, used as a feedstock, or used as refrigerant in appliances manufactured prior to January 1, 2020. Production frozen January 1, 2015 and phased out January 1, 2030. Administrator authorized to grant limited exceptions consistent with the Montreal Protocol. EPA to issue regulations by December 31, 1999.

Accelerated Reduction Schedule

Administrator to promulgate regulations for an accelerated phase-out of Class I or Class II substances if he determines, based on credible scientific information, that acceleration may be necessary to protect human health and environment; if available substitutes make it practicable; or if the Montreal Protocol is modified to require faster reductions. Persons may petition for an accelerated schedule and EPA must grant or deny such petitions in 180 days.

Exchange

A company may produce or import a different mix of substances than it produced or imported in baseline year, or may trade production or consumption allowances with another company, if the change in mix on the trade results in greater total reductions for each substance than would otherwise be achieved. EPA to issue regulation within 10 months of enactment.

Use, Recycling and Disposal

Lowest achievable level of use and emissions, maximum recycling, and safe disposal of Class I substances used as a refrigerant required as of July 1, 1992. Regulations due by January 1, 1992. EPA to similarly regulate all other uses of Class I substances and all Class II substances within 4 years of enactment.

Venting of Class I and Class II substances during the servicing or disposal of refrigeration equipment is prohibited as of July 1, 1992.

Mobile Air Conditioning

Recycling of Class I or Class II substances used in motor vehicle air conditioning required as of January 1, 1992 (for shops servicing fewer than 100 vehicles, January 1, 1993). Recycling equipment and operators must be certified. Regulations due within one year of enactment. Sale of small containers of Class I or Class II substances except to certified mechanics banned within 2 years.

Nonessential Products

CFC-containing party streamers, noise horns, cleaning fluids for noncommercial photo and electronic equipment and other consumer products deemed nonessential by the Administrator banned within 2 years of enactment. Regulations due within one year of enactment.

Effective January 1, 1994, aerosols containing HCFCs and/or plastic foam products made with HCFCs are banned. Exceptions may be granted for aerosols found essential as a result of flammability or worker safety concerns. Foam insulation and rigid foams necessary to meet auto safety standards also exempt.

Labeling

Containers that contain Class I and II substances and products containing Class I substances to be labelled beginning in 30 months after enactment. Products containing or manufactured with Class II substances to be labelled after 30 months from enactment if Administrator finds, after public comment, alternatives are available that reduce overall risk to human health and the environment. Products made with Class I substances must be labelled, 30 months after enactment, unless Administrator finds no alternatives available. Regulations due within 18 months of enactment. Effect 1/15/15/ [*sic*], all products containing Class II substances, or manufactured with Class I or II substances must be labelled.

Safe Alternatives

EPA to take specified actions to facilitate and encourage development of safe substitutes. Regulations also required to make it unlawful to replace any Class I or Class II substance with a [line missing from text] and safety data on substitutes to the Agency and notify EPA 90 days before introducing chemical for significant new use as substitute.

Procurement

Within 18 months, EPA, The General Services Administration, and The Department of Defense to promulgate procurement regulations requiring maximum substitution of safe alternatives for Class I and Class II substances.

Methane

Five reports required in 2 years, one in 4 years, to identify sources of domestic and international methane emissions, potential for preventing increases and options to stop or reduce growth of emissions.

Source: "Clean Air Act Amendments of 1990: Detailed Summary of Titles." Washington, DC: U.S. Environmental Protection Agency, 30 November 1990.

Executive Order 12843 on Procurement of Ozone-Depleting Substances, 1993

On 21 April 1993, President Bill Clinton signed an executive order outlining federal policies on the purchase of ozone-depleting substances. Portions of that executive order are reprinted below.

Whereas, the essential function of the stratospheric ozone layer is shielding the Earth from dangerous ultraviolet radiation; and

Whereas, the production and consumption of substances that cause the depletion of stratospheric ozone are being rapidly phased out on a worldwide basis with the support and encouragement of the United States; and

Whereas, the Montreal Protocol on Substances That Deplete the Ozone Layer, to which the United States is a signatory, calls for a phaseout of the production and consumption of these substances; and

Whereas, the Federal Government, as one of the principal users of these substances is able through affirmative procurement practices to reduce significantly the use of these substances and to provide leadership in their phaseout; and

Whereas, the use of alternative substances and new technologies to replace these ozone-depleting substances may contribute positively to the economic competitiveness on the world market of U.S. manufacturers of these innovative safe alternatives;

Now, Therefore, I, William Jefferson Clinton, by the authority vested in me as President by the Constitution and the laws of the United States of America, including the 1990 amendments to the Clean Air Act ("Clean Air Act Amendments"), Public Law 101-549, and in order to reduce the Federal Government's procurement and use of substances that cause stratospheric ozone depletion, do hereby order as follows:

Section 1. *Federal Agencies.* Federal agencies shall, to the extent practicable:

(a) conform their procurement regulations and practices to the policies and requirements of Title VI of the Clean Air Act Amendments, which deal with stratospheric ozone protection;
(b) maximize the use of safe alternatives to ozone-depleting substances;
(c) evaluate the present and future uses of ozone-depleting substances, including making assessments of existing

and future needs for such materials, and evaluate their use of and plans for recycling;

(d) revise their procurement practices and implement cost-effective programs both to modify specifications and contracts that require the use of ozone-depleting substances and to substitute non–ozone-depleting substances to the extent economically practicable; and

(e) exercise leadership, develop exemplary practices, and disseminate information on successful efforts in phasing out ozone-depleting substances.

Sec. 2. *Definitions*

[This section defines specialized terms used in the executive order.]

Sec. 3. *Policy.* It is the policy of the Federal Government that Federal agencies: (i) implement cost-effective programs to minimize the procurement of materials and substances that contribute to the depletion of atmospheric ozone; and (ii) give preference to the procurement of alternative chemicals, products, and manufacturing processes that reduce overall risks to human health and the environment by lessening the depletion of ozone in the upper atmosphere. In implementing this policy, prior to final promulgation of EPA regulations on Federal procurement, Federal agencies shall begin conforming their procurement policies to the general requirements of Title VI of the Clean Air Act Amendments by:

(a) minimizing, where economically practicable, the procurement of products containing or manufactured with Class I substances in anticipation of the phaseout schedule to be promulgated by EPA for Class I substances, and maximizing the use of safe alternatives. In developing their procurement policies, agencies should be aware of the phaseout schedule for Class II substances;

(b) amending existing contracts, to the extent permitted by law and where practicable, to be consistent with the phaseout schedules for Class I substances. In awarding contracts, agencies should be aware of the phaseout schedule for Class II substances in awarding contracts [*sic*];

(c) implementing policies and practices that recognize the increasingly limited availability of Class I substances as production levels capped by the Montreal Protocol decline until final phaseout. Such practices shall include, but are not limited to:
   (i) reducing emissions and recycling ozone-depleting substances;
   (ii) ceasing the purchase of nonessential products containing or manufactured with ozone-depleting substances; and
   (iii) requiring that new contracts provide that any acquired products containing or manufactured with Class I or Class II substances be labeled in accordance with section 622 of the Clean Air Act Amendments.

Sec. 4. *Responsibilities*

[This section requires agencies to adopt practices that conform to the reduced use of ozone-depleting substances, as outlined in the executive order.]

Sec. 5. *Reporting Requirements*

[This section requires all agencies to report their new procurement procedures to the Office of Management and Budget within six months.]

Sec. 6. *Exceptions*

[This section states the conditions under which exceptions can be granted to the executive order.]

Sec. 7. *Effective Date*

[The effective date of the executive order is given as 30 days after issuance of the order.]

Sec. 8. *Federal Acquisition Regulatory Councils*

[This section explains that various regulatory councils are responsible for seeing that the executive order is properly carried out.]

Sec. 9. *Judicial Review*

[This section outlines the legal protection due the United States government under this order.]

William J. Clinton
The White House
April 21, 1993

National Weather Service Ultraviolet Light Exposure Index

In 1994, the National Weather Service announced that it would be publishing a daily report predicting the amount of exposure to harmful ultraviolet light that could be expected on that day. Data for the exposure index are obtained from satellite observations of the concentration of stratospheric ozone. Readings on the index are shown in Table 10.

TABLE 10 National Weather Service UV Light Exposure Index

		Danger Level for:	
Index	Level	Fair-Skinned People	Darker-Skinned People
0–2	(Minimal)	up to 30 mins	up to two hours
3–4	(Low)	15–20 mins	75–90 mins
5–6	(Moderate)	10–12 mins	50–60 mins
7–9	(High)	7–8-1/2 mins	33–40 mins
10+	(Very High)	4–6 mins	20–30 mins

Source: National Weather Service

Corporate Phaseout Goals

As evidence for the effects of CFCs on stratospheric ozone became stronger, a number of individual corporations and corporate groups began to voluntarily set dates for phasing out their production and use. Table 11 presents a partial list of such organizations presented before a committee hearing of the U.S. House of Representatives in 1990.

TABLE 11 CFC Phaseout Goals of Major Corporations and Corporate Groups Worldwide

Corporation or Corporate Group	Home Country	Target Date
Foodservice and Packaging Institute	US	1988
Siemens Germany	Germany	1989
Northern Telecom	Canada	1991
IBM	US	1993
Seiko Epson	Japan	1993
AT&T	US	1994
Volvo	Sweden	1994
Hoechst Ag	EEC	1995
Polyisocyanurate Insulation Mfrs.	US	1995
Nissan	Japan	1995+
Sharp	Japan	1998
DEC	US	2000
du Pont	US	2000
Hitachi	Japan	2000
Matsushita	Japan	2000
NEC	Japan	2000
Polyurethane Foam Association	US	2000
Toshiba	Japan	2000

Source: "Stratospheric Ozone Depletion," Hearings before the U.S. House of Representatives, Committee on Health and the Environment, January 25, 1990. Washington, DC: U.S. Government Printing Office, p. 786.

Directory of Organizations 5

Nongovernmental Organizations

Alliance for Responsible Atmospheric Policy
2111 Wilson Boulevard, Suite 850
Arlington, VA 22201
(703) 243-0344
(703) 243-2874 (FAX)

Originally organized in 1980 as the Alliance for Responsible CFC Policy, this group is a coalition of companies that manufacture chlorofluorocarbons, hydrochlorofluorocarbons, and hydrofluorocarbons. It works to represent the interest of industry in the development of international and national policies about the production and consumption of potential ozone-depleting chemicals. Its stated goal is to ensure "an appropriate global approach to the issue [of ozone depletion] while minimizing costly and ineffective regulations on user industries."

Publications: A briefing book, *Ozone Protection Policies*, that is regularly updated; periodic bulletins and documents dealing with federal and international legislative and regulatory actions pertaining to CFC phaseout, HCFC controls, global warming, and related issues; periodic newsletters describing local, state, and federal laws.

American Association for Aerosol Research (AAAR)
Kemper Meadow Center
1330 Kemper Meadow Drive
Cincinnati, OH 45240
(513) 742-2227
(513) 742-3355 (FAX)

AAAR was founded in 1982 by a group of professionals concerned about the importance of aerosols in a number of social, political, and environmental areas, including air pollution, atmospheric sciences, industrial hygiene, and nuclear safety. The organization provides a forum in which issues can be addressed and knowledge shared among professionals engaged in aerosol research. Membership is open to anyone who agrees with the organization's goals of advancing scientific and technological knowledge about aerosols and disseminating technical information in the field.

Publication: A scientific journal, *Aerosol Science and Technology.*

Center for Environmental Information
50 West Main Street
Rochester, NY 14614-1218
(716) 262-2870
(716) 262-4156 (FAX)

The Center for Environmental Information is a private, nonprofit organization that collects and disseminates information on environmental issues, including ozone depletion. The center is a local organization that has served the Rochester/Finger Lakes region for 20 years. It maintains a library that is open to the general public and provides bibliographies and other information on environmental issues. The center also sponsors a course in environmental law, hosts local seminars on environmental topics, offers training in dealing with environmental issues, and puts on an annual conference in Washington, D.C., on global change issues.

Publications: *Global Climate Change Digest,* a monthly journal providing current information on greenhouse gases and ozone depletion; *Environmental Events Calendar,* listing activities of various environmental groups; *Directory of Environmental Agencies and*

Organizations, a guide to more than 350 agencies and organizations involved with environmental issues in the Rochester area.

Center for Global Change
University of Maryland at College Park
The Executive Building, Suite 401
7100 Baltimore Avenue
College Park, MD 20740
(301) 403-4165
(301) 403-4292 (FAX)

The Center for Global Change was founded in 1989 with a grant from the Environmental Protection Agency. It was created to provide a bridge among scientists, policy makers, and business leaders on issues of global environmental change such as ozone depletion, global climate change, energy efficiency, and renewable energy. The center seeks innovative solutions to environmental problems and studies their relationship to energy use, economic development, and equity among peoples and nations. It evaluates and recommends policies, technologies, and institutional reforms with the goal of promoting sustainable development with minimal environmental damage. Research results are disseminated through publications, seminars, training, and other activities.

Publications: The center does not have a publishing program per se, but reports on its research in various national publications. Numerous books and articles reporting on research conducted at the center include the following: "Stratospheric Ozone Depletion: Can We Save the Sky?" in *Green Globe Yearbook;* "Tax Tools for Protecting the Atmosphere: The U.S. Ozone-Depleting Chemicals Tax" in *The Greening of Government Taxes and Subsidies: An International Casebook of Leading Practices;* "History of the Montreal Protocol's Ozone Fund" in *International Environmental Reporter;* and "The Development of Substitutes for Chlorofluorocarbons: Public-Private Cooperation and Environmental Policy" in *Ambio.*

Chemical Manufacturers Association
2501 M Street, NW
Washington, DC, 20037
(202) 887-1100

The association was founded in 1872 by manufacturers of chemicals. It supports research for control of air and water pollution, and since 1972 has funded work on the behavior of CFCs in the air. The association provides information to the public through its Chemical Referral Center. It maintains an education program for schools and gives annual awards to chemistry teachers. The association coordinated some of the first research by chemical manufacturers on the interaction of chlorofluorocarbons and the ozone layer.

Publication: A free newsletter, *ChemEcology*, published ten times per year.

Climate Institute
324 Fourth Street, NE
Washington, DC 20002-5821
(202) 547-0104
(202) 547-0111 (FAX)

The Climate Institute is a private, nonprofit organization founded in 1986 to draw closer attention to environmental issues such as global climate change and ozone destruction that had largely been ignored to that point in time. The institute's mission is to promote understanding among concerned citizens, scientists, corporate leaders, and policy makers about such issues. The institute works through ministerial briefings, presidential conferences, international conferences, workshops, and diplomatic symposia in nations around the world.

Publications: A book, *Coping with Climate Change;* reports such as *Report of the International Workshop on Framework Convention and Associated Protocols: A Nongovernmental Perspective* and *The Arctic and Global Change;* a newsletter, *Climate Alert*.

Environmental Defense Fund
257 Park Avenue South
New York, NY 10010
(212) 505-2100
(212) 505-2375 (FAX)

The Environmental Defense Fund is a nonprofit organization founded in 1967 to search for interdisciplinary, innovative, economically viable solutions to major environmental problems. It currently has more than 250,000 members and an annual budget in excess of $22 million. The Fund supports a staff of more than 50 full-time scientists, engineers, lawyers, and economists working on problems such as ozone depletion, acid rain, tropical forest destruction, toxic wastes, global warming, and protection of wildlife and habitats.

Publications: A monthly newsletter, *EDF Letter;* many books, articles, pamphlets, reprints, and other publications, including *Cutting the Costs of Environmental Policy: Lessons from Business Responses to CFC Regulation, The Global Environmental Agenda for the United States and Japan, Global Lessons from the Ozone Hole,* and *Protecting the Ozone Layer: What You Can Do.*

Friends of the Earth
1025 Vermont Avenue NW, Third Floor
Washington, DC 20005
(202) 783-7400
(202) 783-0444 (FAX)

Friends of the Earth was founded in 1969 by David Brower, formerly executive director of the Sierra Club. In its first year, the organization launched three initiatives: challenging environmentally destructive technology, informing citizens about critical issues, and expanding opportunities for ordinary citizens to become involved in environmental decisions. A key issue around which the Friends organized their early activities was opposition to supersonic aircraft, which threatened to damage the ozone layer. Today, the organization emphasizes an interest in taking environmentalism in new directions, pioneering and redefining what environmental advocacy is all about.

Publications: Books, such as *Into the Sunlight: Exposing Methyl Bromide's Threat to the Ozone Layer* and *Environmental Education Resource Guide;* a quarterly newsletter, *Atmosphere;* a fact sheet, *Ozone Campaign.*

Global Foundation, Inc.
1450 Madruga Avenue, Suite 301
Coral Gables, FL 33146
(305) 669-9411
(305) 669-9464 (FAX)

The Global Foundation is a private, nonprofit organization founded to support research and to disseminate information on environmental problems. The foundation sponsors lectures, workshops, and conferences of scientists, industrialists, government officials, and representatives from nongovernmental organizations. It focuses on four major issues of international significance: The

Nuclear Quagmire, International Environmental Problems, Problems of Energy, and Medicine. In addition, the Foundation engages in theoretical research on problems such as elementary particles and the nature of the universe, the origin of the solar system, and the evolution of the genetic code.

Publications: Monographs, white papers, news articles, and proceedings of seminars, workshops, international conferences and international forums.

National Resources Defense Council
1350 New York Avenue, NW
Washington, DC 20005
(202) 783-7800
(202) 783-5917 (FAX)

The National Resources Defense Council was founded in 1970 as a nonprofit organization dedicated to protecting the nation's air, water, and food supplies through legal action, scientific research, advocacy, and public education. One of the council's most effective approaches has been to sue private companies that pollute the environment. Recently it has expanded its efforts to the international level, promoting exchanges between individual scientists and scientific organizations from the United States, the former Soviet Union, and other nations. Since 1988, the council's work on ozone depletion has been carried out as part of its Atmosphere Protection Initiative (API). API is a multidisciplinary effort to address the related problems of global warming, ozone depletion, acid rain, and urban smog.

Publications: A quarterly, *The Amicus Journal,* and a newsletter, *NRDC Newsline,* published five times a year. Various books and pamphlets on a wide variety of topics, including *A Who's Who of American Ozone Depleters: A Guide to 3,014 Factories Emitting Three Ozone-Depleting Chemicals* and *Public Enemy No. 1,1,1: A Who's Who of Consumer Products That Contain the Ozone-Depleting Chemical 1,1,1-Trichloroethane.*

Resources for the Future
1616 P Street
Washington, DC 20036
(202) 328-5000

Resources for the Future (RFF) is an independent, nonprofit organization founded in 1952 through a grant from the Ford

Foundation. The purpose of the organization is to inform and improve policy debates about issues relating to natural resources and the environment. The work of RFF is carried out through three units, the Energy and Natural Resources Division, the Quality of the Environment Division, and the Center for Risk Management. RFF sponsors research, carries out educational programs, holds conferences and other types of meetings, provides grants and fellowships, and arranges policy briefings for legislators, business leaders, and the media. The organization does not, itself, take stances on any side of any particular issue, but maintains a nonpartisan position in all of its work.

Publications: A quarterly periodical, *Resources;* a quarterly summary of research activities, *RFF Research Digest;* policy briefs; discussion papers; and more than 100 books and studies distributed by Johns Hopkins Press.

Worldwatch Institute
1776 Massachusetts Avenue, NW
Washington, DC 20036-1904
(202) 452-1999
(202) 296-7365 (FAX)

The Worldwatch Institute is a nonprofit, public interest research organization whose primary interests are global environmental and environmentally related issues. The institute was founded in 1974 in an effort to raise public awareness about such issues so that citizens will be more willing to support policy initiatives that deal with environmental concerns. One of the most important characteristics of Worldwatch is its interdisciplinary approach to problems in which it is interested. As an example, the institute's annual State of the World attempts to provide a single report on the status of population, environment, agriculture, food, economic and social conditions, and other relevant aspects of human societies. Research sponsored and carried out by the institute is disseminated in a variety of ways, including books, articles, television and radio interviews, talks and lectures, and testimony before legislative bodies throughout the world.

Publications: An annual report, *State of the World;* a bimonthly magazine, *World Watch;* an annual handbook of statistics, *Vital Signs: The Trends That Are Shaping Our Future; World Watch Magazine; The World Watch Reader on Global Environmental Issues;* and more than 100 *Worldwatch Papers* on topics such as ozone depletion, global warming, biological diversity, poverty and the environment, the green revolution, and reforesting the Earth.

Governmental Agencies

Atmospheric Environment Service
Environment Canada
4905 Dufferin Street
Toronto, Ontario, M3H 5T4
Canada
(800) 668-6767

The Atmospheric Environment Service is a division of the Canadian government's primary agency for dealing with environmental issues, Environment Canada. The service is the primary means by which the government attempts to inform citizens about environmental issues relating to the atmosphere, such as ozone depletion and global warming. In addition to producing a wide variety of publications of exceptional quality, the service is active in promoting Environmental Citizenship, a program created by Canada's Green Plan to create a society in which individuals and groups have the knowledge and understanding that will lead to responsible environmental action.

Publications: A variety of reports, books, and pamphlets on many aspects of the atmosphere, including *The Ozone Layer: What's Going on up There?; UV and You: Living with Ultraviolet; Replenishing the Ozone Layer* (an Update on Environmental Issues); *Stratospheric Ozone Depletion* (an Environmental Indicator Bulletin); *Guarding Our Earth: Ozone; The Ozone Layer* (a Fact Sheet); *Canada's Ozone Layer Protection Program—A Summary;* and *A Primer on Ozone Depletion.*

International Union of Geodesy and Geophysics
Observatoire Royal
Avenue Circulaire 3
B-1180, Brussels, Belgium

The IUGG was founded in 1919 to promote scientific studies of the Earth's environment, including the atmosphere and the Sun-Earth relationship. It sponsored the first international ozone research conference in 1929. The IUGG presently represents 79 countries and is composed of semi-independent associations, of which two are directly concerned with ozone depletion. The International Association of Geomagnetism and Aeronomy conducts Antarctic research and mid-atmospheric research. The International Association of Meteorology and Atmospheric Physics performs studies of the upper atmosphere's weather and pollution, as well as of ozone and climate. It also supervises the International Ozone Commission, established in 1948 to coordinate ozone research.

National Academy of Sciences
National Research Council
2101 Constitution Avenue, NW
Washington, DC 20418
(202) 334-2000

Established by act of Congress in 1863 to advise government on scientific and technological issues, the National Academy of Sciences is one of the most prestigious scientific organizations in the United States. In 1916, the Academy established the National Resource Council (NRC) to coordinate research in government, higher education, and industries. Some of the most important studies of ozone depletion have been produced by the NRC.

Publications: A number of research reports, including *Environmental Impact of Stratospheric Flight; Ozone Depletion, Greenhouse Gases, and Climate Change; Atmospheric Ozone Studies: An Outline for an International Observation Program; Causes and Effects of Stratospheric Ozone Reduction: An Update; Causes and Effects of Changes in Stratospheric Ozone: Update 1983; Halocarbons: Effects on Stratospheric Ozone;* and *Protection against Depletion of Stratospheric Ozone by Chlorofluorocarbons*.

United Nations Environment Programme
Ozone Secretariat
P.O. Box 30552
Nairobi, Kenya
(254-2) 623-885
(254-2) 521-930 (FAX)

North America Regional Office
UNEP Liaison Office for UNEP
UNDC Two Building, Room 0803
Two United Nations Plaza
New York, NY 10017
(212) 963-8138

Washington, DC, Liaison Office
Ground Floor
1889 F Street, NW
Washington, DC 20006
(202) 289-8456

The United Nations Environment Programme (UNEP) was established in December 1972 as a result of the United Nations Conference on the Environment held in Stockholm, Sweden, earlier that year. UNEP promotes international cooperation on the environment and the use of the world's resources. It has developed a number of educational programs and draws up action plans for regions and countries. Its Coordinating Committee on the Ozone Layer was responsible for implementing the 1977 World Plan of Action on the Ozone Layer and for laying the groundwork for the 1985 Geneva Convention that led to the Montreal Protocol. UNEP is now home to the Ozone Secretariat, established as a result of the Protocol agreement.

Publications: UNEP offers many popular and technical publications, including *Handbook for the Montreal Protocol on Substances That Deplete the Ozone Layer; Report of the Halons Technical Options Committee; Protecting the Ozone Layer; Scientific Assessment of Ozone Depletion: 1991; The Impact of Ozone Layer Depletion; Ozone Depletion: Implications for the Tropics;* and *OzonAction*, a quarterly newsletter.

U.S. Environmental Protection Agency
401 M Street, S.W.
Washington, DC 20460
(202) 260-2080

The U.S. Environmental Protection Agency was created by an act of Congress in 1970 to coordinate efforts of the federal government in protecting and enhancing the environment. It carries out activities to control and reduce pollution in the areas of air, water, solid waste, pesticides, radiation, and toxic substances. It sponsors and coordinates research and antipollution activities by federal, state, and local groups, public and private organizations, and individuals. It is responsible for setting standards, monitoring the environment, and enforcing laws and regulations dealing with the environment. The EPA is responsible for carrying out provisions of the 1990 Clean Air Act regarding protection of the ozone layer in the United States.

Publications: A monthly magazine, *EPA Journal,* and many reports and educational publications on environmental issues, including ozone depletion, such as *The Plain English Guide to the Clean Air Act; Protecting the Ozone Layer: A Checklist for Citizen Action;* and *Auto Air Conditioners and the Ozone Layer: A Consumer Guide.* The agency also operates a Stratospheric Ozone Information Hotline providing information on production phaseout and controls, alternative substances, product labeling, federal procurement, and other topics.

U.S. National Aeronautics and Space Administration
Office of Space Science and Applications
Earth Science and Applications Division
Two Independence Square
300 E Street, SW
Washington, DC 20546
(202) 453-1000

Goddard Space Flight Center
Greenbelt, MD 20771
(301) 286-2000
(301) 286-1707 (FAX)

NASA was established by an act of Congress in 1958 to conduct research on problems of flight within and outside the Earth's atmosphere. Although its primary goals have centered on the development of manned and unmanned space flight systems, it has also pursued an active program of research on near-Earth flight. Its Goddard Space Flight Center has been an important center of research on chemistry of the stratosphere and ozone layer depletion.

Publications: Many technical reports on ozone depletion, often in association with other agencies, including *Scientific Assessment of Ozone Depletion: 1991; The Atmospheric Effects of Stratospheric Aircraft: A Topical Review; Two-Dimensional Intercomparison of Stratospheric Models; Study of High Speed Civil Transport;* and *The Atmospheric Effects of Stratospheric Aircraft: A First Program Report.*

U.S. National Oceanic and Atmospheric Administration
Office of Global Programs
1100 Wayne Avenue, Suite 1225
Silver Spring, MD 20910
(301) 427-2089
(301) 427-2082 (FAX); (301) 427-2073 (FAX)

Aeronomy Laboratory
325 Broadway
Boulder, CO 80303-3328
(303) 497-5785
(303) 497-5373 (FAX)

One of the lead agencies in the United States for research on ozone depletion is the National Oceanic and Atmospheric Administration (NOAA), through its Office of Global Programs. A major center for NOAA's research on ozone depletion is the Aeronomy Laboratory at the Environmental Research Laboratories in Boulder, Colorado. In addition to ozone depletion, the laboratory works on acid deposition, tropospheric ozone production, tropical ocean–atmosphere interactions, and the greenhouse effect. The laboratory is staffed with approximately 100 scientists, engineers, and support personnel.

Publications: Numerous technical reports and a public information booklet, *Our Ozone Shield,* second in a series of *Reports to the Nation*, prepared jointly by the Office of Global Programs and the University Corporation for Atmospheric Research Office for Interdisciplinary Studies.

World Meteorological Organization
Ozone Secretariat
41 Avenue Giuseppi Mota
P. O. Box 2300
Geneva, 20, CH 1211
Switzerland
41-22-730-81-11
41-22-734-23-26 (FAX)

The World Meteorological Organization (WMO) is the successor to the International Meteorological Organization (IMO), originally founded in 1873 to coordinate research on weather and climate around the world. In 1947, officials of the IMO convened to consider plans for reorganizing the organization under the auspices of the United Nations. As a result of that meeting, the new World Meteorological Organization was established and began operation in 1951. The policies under which WMO operates are established at quadrennial meetings of the World Meteorological Congress. Between these sessions, policies are implemented by a 29-member Executive Committee. The organization's secretariat is located in Geneva. The work of the WMO is carried out through special commissions dealing with issues such as climate, the oceans, and ozone layer depletion, and through regional associations in which members focus on issues that are particularly relevant to their own areas of the world. WMO has taken an activist stance on the question of ozone depletion for more than two decades. In addition to the scientific and technological research it supports on ozone depletion, WMO works with the United Nations Environment Programme to study the social and economic effects of ozone-depleting chemicals and to encourage efforts to reduce human dependence on these materials.

Publications: Many technical reports and other publications, including *Atmospheric Ozone, 1985: Assessment of Our Understanding of the Processes Controlling Its Present Distribution and Change; Scientific Assessment of Stratospheric Ozone: 1989;* and *Scientific Assessment of Ozone Depletion: 1991.*

Print Resources

This chapter includes bibliographic citations and annotations of relevant print resources, including a number of technical and popular books on ozone depletion. It also includes a listing of reports of hearings held by relevant committees of the U.S. Senate and House of Representatives.

Bibliography

The Greenhouse Effect: A Bibliography. Santa Cruz, CA: Reference and Research Services, 1990. 60pp. ISBN 0-937855-34-0.

In spite of its title, this little book contains a number of references dealing with ozone, CFCs, and related topics. It is now somewhat out of date, but contains enough early references to make it a valuable addition to the library.

Books

Benedick, Richard E. ***Ozone Diplomacy: New Directions in Safeguarding the Planet.*** Cambridge, MA: Harvard University Press, 1991. 300pp. ISBN 0-674-65000-X.

Benedick was leader of the U.S. negotiating team for the Montreal Protocol. His book describes the process by which agreement was reached on the protocol and suggests that that process and the protocol itself might serve as a model for negotiations over global environmental issues such as global warming. A particularly valuable aspect of the book is that it includes the complete texts of the Vienna, Montreal, and London meetings on ozone depletion.

Biggs, R. Hilton, and Margaret E. B. Joyner, eds. ***Stratospheric Ozone Depletion/UV-B Radiation in the Biosphere***. Berlin: Springer-Verlag, 1994. 358pp. ISBN 0-387-57810-2.

This book contains papers presented at the NATO Advanced Research Workshop on Stratospheric Ozone Depletion/UV-B Radiation in the Biosphere that was held at Gainesville, Florida, from 14 to 18 June 1993. The papers are quite technical, but of considerable interest to specialists in the field.

Biswas, Asit K. ed. ***The Ozone Layer.*** Oxford: Pergamon Press, 1979. 381pp. ISBN 0-08-022429-6.

This volume is a collection of papers presented at a meeting of experts designated by governments and by intergovernmental and nongovernmental organizations, organized by the United Nations Environment Programme and held in Washington, DC, from 1 to 9 March 1977. The papers cover a wide range of topics, including Aircraft Engine Emissions, Effects of Ultraviolet Radiation on Human Health, Ozone Layer and Development, A Summary of the Research Program on the Effect of Fluorocarbons on the Atmosphere, Atmospheric Exchange Processes and the Ozone Problem, Stratospheric Ozone Depletion—An Environmental Impact Assessment, and U.S. Investigations to Evaluate the Potential Threat of Stratospheric Ozone Diminution. The last section of the book contains the World Plan of Action, an agenda adopted by members of the convention.

Bojkov, Rumen D., and Peter Fabian, eds. ***Ozone in the Atmosphere: Proceedings of the Quadrennial Ozone Symposium and Tropospheric Ozone Workshop.*** Hampton, VA: A. Deepak, 1989. 822pp. ISBN 0-937194-15-8.

The 192 scientific reports in this volume were presented at the sixteenth quadrennial symposium organized by the International

Ozone Commission. The papers are divided into ten major sections: Polar Ozone, Analysis of Ozone Observations, Ozone Observations from Satellites, Observations of Trace Constituents Relevant to the Ozone Issue, Ozone and Radiation, Solar Effects and Mesosphere, Tropospheric Ozone, Chemical-Radiative-Dynamical Model Studies, Modeling, Reaction Kinetics and Laboratory Studies, and Developments in Observational Techniques. Although the reports are very technical, a number of graphs are easily understood by the layperson and provide stark illustration of ozone depletion patterns in various parts of the planet.

Bower, Frank A., and Richard B. Ward, eds. ***Stratospheric Ozone and Man.*** 2 vols. Boca Raton, FL: CFC Press, 1981. 217pp. and 261pp. ISBNs 0-8493-5753-5 and 0-8493-5755-1.

The editors explain that this book was originally intended as a survey of the scientific and technological aspects of the ozone depletion question. As the work developed, however, it became clear that the science and technology could not be separated from policy issues. As a result, the final text contains detailed discussions of many facets of the ozone depletion issue. The two volumes are divided into five parts: Stratospheric Ozone, Atmospheric Processes Influencing Stratospheric Ozone, Does Man Influence Stratospheric Ozone?, Effects and Research, and Public Policy.

Brown, Lester R., et al. ***State of the World.*** New York: W. W. Norton, published annually. ISBN 0-393-03717-7 (1995 edition).

State of the World, published annually, presents an evaluation of major environmental issues throughout the world. It is an excellent source of up-to-date statistics on topics such as ozone depletion, global warming, population issues, pollution, and the like. In spite of its tendency to overstate risks in some cases, it is an enormously valuable resource.

Brunnee, Jutta. ***Acid Rain and Ozone Layer Depletion: International Law and Regulation.*** Dobbs Ferry, NY: Transnational Publishers, 1988. 302pp. ISBN 0-9413-2051-0.

This book deals with problems surrounding the regulation of acid rain and ozone layer depletion in many countries around the world. It discusses some of the issues involved in obtaining international cooperation in the management of air quality.

Cagin, Seth, and Philip Dray. ***Between Earth and Sky: How CFCs Changed Our World and Endangered the Ozone Layer.*** New York: Pantheon Books, 1993. 430pp. ISBN 0-679-42052-5.

One of the most thoroughly researched, enjoyably written, and carefully balanced presentations of the story of ozone depletion and the special role played by CFCs in that phenomenon. The authors provide an excellent historical background about ozone-depleting chemicals and the discoveries that led to the realization of an ozone hole. Of particular interest is the insightful analysis of the political issues involved in solving the ozone controversy. The book is complemented by a very complete set of research notes, an extensive bibliography, and an unusually thorough index. One of the most highly recommended books on the subject now available.

Caldicott, Helen. ***If You Love This Planet: A Plan To Heal the Earth.*** New York: W. W. Norton, 1992. 231pp. ISBN 0-393-30835-9 (pbk).

Dr. Caldicott is an Australian physician who was one of the founders of Physicians for Social Responsibility. This book deals with a wide range of environmental problems, including ozone depletion, global warming, overpopulation, species extinction, and toxic pollution. Caldicott shows how these environmental issues are related to social and political policies and structures in the United States and throughout the world.

Climate Impact Committee. ***Environmental Impact of Stratospheric Flight.*** Washington, DC: National Academy of Sciences, 1975. 348pp. ISBN 0-309-02346-7.

This report summarizes the work of a joint committee of the National Research Council, the National Academy of Sciences, and the National Academy of Engineering. The report considers the biological and climatic effects of aircraft emissions on the stratosphere. The report was largely motivated by predictions that the use of SSTs could have a devastating effect on the ozone layer in the stratosphere. Authors of the report concluded that an increase in the number of airplanes traveling in the stratosphere could diminish the amount of ozone with a consequent increase in the intensity of ultraviolet light reaching the ground. They warned that this sequence of events could very well result in a number of harmful effects to organisms on the Earth's surface.

The Crucial Decade: The 1990s and the Global Environmental Challenge. Washington, DC: World Resources Institute, 1989. 30pp. ISBN 0-915825-37-6.

This short book brings together descriptions of critical world environmental problems, such as global warming and ozone depletion, and provides a checklist of ways to deal with such issues. The book argues that concerted efforts can and must be made to deal with these problems. The special role of population growth is described as a critical factor in the development of these problems.

Cumberland, John H., James R. Hibbs, and Irving Hoch, eds. ***The Economics of Managing Chlorofluorocarbons: Stratospheric Ozone and Climate Issues.*** Washington, DC: Resources for the Future, 1982. 511pp. ISBN 0-8018-2963-1.

A strictly economic approach to the issue of ozone depletion is taken in this book. Authors of various chapters discuss issues such as Costs and Benefits of Chlorofluorocarbon Control; Climate, Energy Use, and Wages; Environmental Standards and the Management of CFC Emissions; Controlling Refrigerant Uses of Chlorofluorocarbons; and Impacts of Ozone Reduction on National and International Commercial Fisheries.

Dobson, Gordon M. B. ***Exploring the Atmosphere.*** Oxford: Clarendon Press, 1963. 209pp. No ISBN.

Dobson is regarded as one of the pioneers of research on the stratosphere, particularly on the ozone layer. This book is, therefore, largely of historical interest. His chapter on the ozone layer is, of course, of greatest interest to modern readers. It includes a striking figure (on page 116) showing how the presence of ozone in the stratosphere effects the level of ultraviolet radiation reaching the Earth's surface. The book is also of some interest because modern critics of the CFC–ozone depletion theory rely heavily on what Dobson wrote three decades ago (while largely ignoring the vast amount of research that has gone on since his time).

Dotto, Lydia, and Harold Schiff. ***The Ozone War.*** Garden City, NY: Doubleday, 1978. 342pp. ISBN 0-385-12927-0.

This book is particularly interesting since it was written so early in the current controversy about ozone depletion. A number of major discoveries had not yet been made, and the authors are quite correct in approaching the subject with a great deal more ambiguity than is found in later works. The authors, one a science writer and the other a professor of chemistry, begin by providing a good summary of the scientific information regarding ozone depletion. They then outline the political and economic factors involved in decisions that had been made to that time and in others that might have to be made in the future.

Environmental Protection Agency. ***Effects of Changes in Stratospheric Ozone and Global Climate.*** Vol. I, Overview, 1986. Vol. II, Stratospheric Ozone, 1986. Vol. III, Assessing the Risks of Trace Gases That Can Modify the Stratosphere, 1987. Washington, DC: Government Printing Office, 1986–1987.

———. ***CFCs and Stratospheric Ozone.*** Washington, DC: Government Printing Office, 1987.

———. ***How Industry Is Reducing Dependence on Ozone-Depleting Chemicals.*** Washington, DC: Government Printing Office, 1988.

———. ***Regulatory Impact Analysis: Protection of Stratospheric Ozone.*** Washington, DC: Government Printing Office, 1988.

The volumes cited above are summaries of the Environmental Protection Agency's major research efforts in the field of ozone depletion. Each provides some of the most reliable and up-to-date (as of the publication date) information on the topic described in the title.

Firor, John. ***The Changing Atmosphere: A Global Challenge.*** New Haven, CT: Yale University Press, 1990. 145pp. ISBN 0-300-03381-8.

The author's thesis is that uncontrolled human population growth is the basis for a number of related atmospheric problems, including acid rain, global warming, and ozone depletion. He suggests that humans will soon have to decide whether to live in the degraded environment that will inevitably result from these problems or whether they will take aggressive actions to reduce population growth and the environmental problems it has produced.

Fisher, David E. ***Fire and Ice: The Greenhouse Effect, Ozone Depletion, and Nuclear Winter.*** New York: Harper & Row, 1990. 232pp. ISBN 0-06-0126214-7.

Fisher's book consists of two parts, the first of which shows the scientific connection among the greenhouse effect, ozone depletion, and nuclear winter. The last of these, nuclear winter, is a "worst case" scenario imagined by scientists in the 1970s and 1980s as a possible result of all-out nuclear war. Theoretical studies of such an event have provided insights into the kinds of environmental changes that might be expected as the result of serious global warming or ozone depletion. The second part of Fisher's book provides a review of some possible solutions for global climate change and ozone depletion.

Fishman, Jack, and Robert Kalish. ***Global Alert: The Ozone Pollution Crisis.*** New York: Plenum Press, 1990. 311pp. ISBN 0-306-43455-5.

Fishman and Kalish provide an interesting, sound, and readable discussion of a number of environmental issues, showing how they are all related to each other. Among the topics included are global warming, tropical deforestation, pollution of the troposphere, and ozone depletion. The authors provide an excellent scientific background for these topics along with a solid introduction to the political, social, and economic issues related to them.

French, Hilary F. ***After the Earth Summit: The Future of Environmental Governance.*** Paper 107. Washington, DC: Worldwatch Institute, 1992. 62pp. ISBN 1-87807-108-4.

This report outlines the general issues faced by the nations of the world in dealing with a number of critical and worldwide environmental problems. The author suggests that the approaches taken thus far, as illustrated by existing treaties on environmental issues, will not be adequate in the future for dealing with potential disasters such as ozone depletion.

Fully Halogenated Chlorofluorocarbons. Geneva: World Health Organization, 1990. 164pp. ISBN 92-4-157113-6.

A publication in the World Health Organization's series on Environmental Health Criteria, this book summarizes a vast amount of fundamental data about CFCs and reviews current information on health effects resulting from exposure to them. The book is a superb, if somewhat dated, reference for basic information on the properties, sources, environmental transport, and effects on plants, animals, and humans of CFCs.

Gore, Al. ***Earth in the Balance: Ecology and the Human Spirit.*** Boston: Houghton Mifflin, 1992. 407pp. ISBN 0-395-57821-3.

Al Gore has made environmental issues a particular concern of his. This book outlines the science and technology as well as the moral, ethical, social, and philosophical issues involved in global environmental issues such as global warming and ozone depletion. The book was written while he was a senator; since his election as vice president, he has served as point man for the Clinton administration on environmental issues.

Gribbin, John. ***The Hole in the Sky: Man's Threat to the Ozone Layer.*** New York: Bantam, 1988.

This book, by a British physicist and experienced science writer, came out just after the Montreal Protocol was signed in 1987 and presents a readable overview of the scientific, political, and diplomatic efforts that resulted in the Protocol. As a scientist, Gribbin knows many of the leading scientists involved in stratospheric ozone research, and he presents their work in a friendly style. The book contains a thorough history of concerns about the ozone layer, including the first ozone layer scare in the late 1960s and early 1970s. The author details the nitrogen cycle and its relation to the ozone layer. He explains how the burning of fossil fuels and the use of agricultural fertilizers add large amounts of nitrogen to the atmosphere and their stratospheric reaction with ozone. He also shows how the 1982 eruption of the El Chichón volcano added ozone-depleting chemicals to the environment and outlines the harmful effects of chlorofluorocarbons.

Horvath, M., L. Bilitzky, and J. Huttner. ***Ozone.*** New York: Elsevier, 1985. 330pp. ISBN 0-444-99625-7.

This is a highly technical text that provides an extensive survey of information currently available on ozone. Included are physical, chemical, and physiological properties of the gas, analytic methods for detecting the presence of ozone and determining ozone levels, techniques for preparing and storing ozone, and uses of the gas. The book is monograph 20 in Elsevier's series on Topics in Inorganic and General Chemistry.

Jones, Robin Russell, and Tom Wigley. ***Ozone Depletion: Health and Environmental Consequences.*** New York: Wiley, 1989. 280pp. ISBN 0-471-92316-8.

The papers in this volume were presented at a conference at the Royal Institute in London in 1988. They deal with both ozone depletion and global warming. Some of the titles included are "Numerical Modeling of Ozone Perturbations," "Present State of Knowledge of the Ozone Layer," "The Montreal Protocol on Substances That Deplete the Ozone Layer: Its Development and Likely Impact," "Epidemiology of Melanoma: Its Relationship to Ultraviolet Radiation and Ozone Depletion," "Consequences for Human Health of Stratospheric Ozone Depletion," and "Ozone Depletion: Consumer Choice."

Maduro, Rogelio A., and Ralf Schauerhammer. ***The Holes in the Ozone Scare: The Scientific Evidence That the Sky Isn't Falling.*** Washington, DC: 21st Century Science Associates, 1992. 356pp. ISBN 0-9628134-0-0.

This book is extremely important since it is probably the best known and most widely quoted text aimed at debunking the concept of ozone depletion and the deleterious effects of CFCs and other so-called ozone-depleting chemicals. The book has been used as a primary reference by Dixy Lee Ray, Rush Limbaugh, and others who dispute the reality of ozone depletion and the effects of CFCs. Interestingly enough, one of the strongest recommendations for the book has come from Lewis du Pont Smith, "heir to the du Pont industrial legacy" and closely associated with the company that was at one time the world's number one producer of CFCs. A particularly interesting feature of the book is the inclusion of Chapter 6 of Gordon Dobson's 1968 text, *Exploring the Atmosphere.*

Makhijani, Arjun, Kevin Gurney, and Annie Makhijani. ***Saving Our Skins: The Causes and Consequences of Ozone Layer Depletion and Policies for Its Restoration and Protection.*** Washington, DC: Institute for Energy and Environmental Research, 1992. 42pp. No ISBN.

This report focuses on the potential role of HCFCs as replacements for CFCs in many industrial and commercial operations.

The authors argue that HCFCs are also damaging to the ozone layer and must be eliminated in the same way that CFCs have been.

McCormick, M. P., and J. E. Lovill, eds. ***Space Observations of Aerosols and Ozone.*** Oxford: Pergamon Press, 1983. 213pp. ISBN 0-08-030427-3.

This volume contains papers presented at a meeting of the Interdisciplinary Scientific Commission of the Committee on Space Research of the International Council of Scientific Unions in Ottawa from 16 May–2 June 1982. Some of the topics covered in the meeting include the Role of Aerosols in Climate, Instrumentation and Future Measurement Systems, Effects on Remote Sensing, and Ozone Variability in Middle Atmosphere.

McCuen, Gary E., ed. ***Our Endangered Atmosphere: Global Warming & the Ozone Layer.*** Hudson, WI: Gary E. McCuen Publications, 1987. 133pp. ISBN 0-86596-063-1.

This book is an anthology of articles dealing with global warming and depletion of the ozone layer. The articles included present more than one side of each issue, some arguing, for example, that neither global warming nor ozone depletion is an important issue. The book, designed for stimulating debate among students, contains useful information for the general reader.

Mészáros, Ernö. ***Global and Regional Changes in Atmospheric Composition.*** Boca Raton, FL: Lewis, 1993. 175pp. ISBN 0-87371-662-0.

The author indicates that his desire in writing this book has been to write "a coherent text, understandable for students in meteorology, chemistry, and ecology, and beginners in the subject." The book describes the composition of the Earth's atmosphere and explains how human activities have brought about changes in that composition. He also describes some scenarios that can be predicted based on the kinds of anthropogenic changes that have already taken place.

Miller, Alan S., and Irving M. Mintzer. ***The Sky Is the Limit: Strategies for Protecting the Ozone Layer.*** Washington, DC: World Resources Institute, 1986. 38pp. ISBN 0-915825-17-1.

Although now seriously dated, this report (Research Report #3 from the World Resources Institute) is one of the most valuable collections of data on the ozone depletion problem. After an introductory section explaining the hypothesized effects of CFCs on the ozone layer, the authors discuss the use and control of CFCs and possible substitutes for the chemicals. In the final section, they outline some fundamental policy issues relating to the regulation of CFCs and other ozone-depleting chemicals.

Mitchell, George J. ***World on Fire: Saving an Endangered Earth.*** New York: Charles Scribner's Sons, 1991. 247pp. ISBN 0-684-19231-4.

Mitchell was majority leader of the U.S. Senate when he wrote this book. He writes clearly about the scientific issues involved in a number of global problems, including depletion of the ozone layer, and he outlines a number of actions that can be taken to deal with these problems.

Myers, Norman. ***Not Far Afield: U.S. Interests and the Global Environment.*** Washington, DC: World Resources Institute, 1987. 84pp. ISBN 0-915825-24-4.

The author argues that it is in the interest of the United States to provide leadership in solving a number of the world's environmental issues, including ozone depletion. The book was written just as the magnitude of that problem was becoming apparent, so its major value is in general principles suggested by the author for dealing with such issues.

Nance, John J. ***What Goes Up: The Global Assault on Our Atmosphere.*** New York: Morrow, 1991. 324pp. ISBN 0-688-08952-6.

Nance explains that his book "is designed to demystify, in layman's terms, the ozone and global-warming issues, and to give you the real story behind the conflicting, competing headlines and news specials." The book leans strongly on the author's interviews with key figures in the ozone depletion and global warming stories: Anderson, Farman, Hansen, Molina, Rowland, Schneider, Solomon, and Watson, among others. The story is easy to follow and is complemented by a bibliography and comprehensive notes.

National Academy of Sciences. ***Halocarbons: Effects on Stratospheric Ozone.*** Washington, DC: National Academy Press, 1976. 352pp. ISBN 0-309-02532-X.

This report by the Panel on Atmospheric Chemistry of the Assembly of Mathematical and Physical Sciences of the National Research Council was one of the earliest and most complete studies on potential effects on atmospheric ozone resulting from aircraft emissions.

———. ***Halocarbons: Environmental Effects of Chlorofluoromethane Release.*** Washington, DC: National Academy Press, 1976. 125pp. ISBN 0-309-02529-X.

A companion volume to the work immediately preceding (*Halocarbons: Effects on Stratospheric Ozone*), this volume reports the findings of the Committee on Impacts of Stratospheric Change of the Assembly of Mathematical and Physical Sciences of the National Research Council.

———. ***Ozone Depletion, Greenhouse Gases, and Climate Change.*** Washington, DC: National Academy Press, 1989. 122pp. ISBN 0-309-03945-2.

This book reports the proceedings of a joint symposium by the Board on Atmospheric Sciences and Climate and the Committee on Global Change of the Commission on Physical Sciences, Mathematics and Resources of the National Research Council held on 23 March 1988. The eleven chapters of the book include presentations by scholars working at the forefront of their fields, including Daniel L. Albritton, F. Sherwood Rowland, Mario J. Molina, and James G. Anderson. The major conclusion of the symposium was that ozone depletion and global warming have to be studied "as parts of an interactive global system rather than as more or less unconnected events."

National Aeronautics and Space Administration. ***Atmospheric Ozone, 1985: Assessment of Our Understanding of the Processes Controlling Its Present Distribution and Change.*** 3 vols. Washington, DC: National Aeronautics and Space Administration, 1985. 1,095 + R86pp. No ISBN.

This large work summarizes all that scientists knew about stratospheric ozone as of 1985. The report was sponsored by a coalition of national and international groups including NASA, the Federal Aviation Administration, the National Oceanic and Atmospheric Administration, the United Nations Environment Programme, the World Meteorological Organization, the Commission of the

European Communities, and the Bundesministerium für Forschung und Technologies of West Germany. Typical of the topics included are Ozone and Temperature Correlations, Ozone and Solar Proton Events, Hydroxyl (OH) Measurements, Methane (CH_4) Measurements, Satellite Data, Chlorine Nitrate, and Interannual Variability.

National Research Council. ***Causes and Effects of Changes in Stratospheric Ozone: Update 1983.*** Washington, DC: National Academy Press, 1984. 254pp. ISBN 0-309-03443-4.

———. ***Causes and Effects of Stratospheric Ozone Reduction: An Update.*** Washington, DC: National Academy Press, 1982. 339pp. ISBN 0-309-03248-2.

———. ***Protection against Depletion of Stratospheric Ozone by Chlorofluorocarbons.*** Washington, DC: National Academy of Sciences, 1979. 392pp. ISBN 0-309-02947-3.

These four volumes contain reports of studies conducted by the National Research Council on various aspects of the ozone depletion issue. The trend across the four reports tends to be an increasing acceptance of the relationship between CFCs and other chemicals and ozone depletion, along with a somewhat diminished estimate of the environmental effect of ozone loss. See Chapter Two for more details on these reports.

Peterson, Elaine. ***Poof: The Aerosol Game.*** New York: Carlton Press, 1989. 80pp. ISBN 0-8062-3256-0.

This short book tells the history of the development of the aerosol industry, with particular emphasis on the role of Harry Edward Peterson (the author's husband) in that development. The problems of ozone depletion are relegated to a short discussion at the end of the book, in which the author complains that the aerosol industry was made a scapegoat for those problems.

Ray, Dixy Lee, with Louis Guzzo. ***Environmental Overkill: Whatever Happened to Common Sense?*** Washington, DC: Regnery Gateway, 1993. 260pp. ISBN 0-89526-512-5.

Ray and Guzzo follow up their earlier book, *Trashing the Planet,* with this aggressive attack on a number of environmental "crises" that they claim have been manufactured by the media, politicians, industrialists, environmentalists, and others with something to gain by promoting nonexistent problems. The authors' case against ozone depletion is weakened to some extent by their use of incomplete and inaccurate scientific information. This book has been used by critics with less scientific background than Ray (such as Rush Limbaugh) to argue against ozone depletion and its association with CFCs.

———. ***Trashing the Planet: How Science Can Help Us Deal with Acid Rain, Depletion of the Ozone, and Nuclear Waste (Among Other Things).*** Washington, DC: Regnery Gateway, 1990. 206pp. ISBN 0-89526-544-3.

The author was long a spokesperson for the position that many environmental issues are grossly exaggerated. She makes this position clear in her introduction to the book, where she states that her purpose is to explain to "sensible citizens . . . what all the environmental fuss is about but whose access to facts is limited to the hyperbole of the popular media."

Reck, Ruth A., and John R. Hummel, eds. ***Interpretation of Climate and Photochemical Models, Ozone and Temperature Measurements.*** New York: American Institute of Physics, 1982. 308pp. ISBN 0-88318-181-9.

This book summarizes a conference on climate modeling held at La Jolla Institute in 1981. The editors point out that this area of modeling is "a particularly challenging specialty within the model-maker's craft." The papers are, in general, very technical, but the general reader may obtain some sense of the kind of work done by modelers and the kinds of results they had obtained in the early 1980s.

Roan, Sharon. ***Ozone Crisis: The 15-Year Evolution of a Sudden Global Emergency.*** New York: Wiley, 1989. 270pp. ISBN 0-471-61985-X.

The author provides a very complete historical account of the development of the debate over ozone depletion beginning with the work of Molina and Rowland in 1973. Each chapter of the book deals with a particular time period in the development of public awareness about this issue. The book also provides a detailed chronology of events between December 1973 and March 1989.

Shea, Cynthia Pollack. ***Protecting Life on Earth: Steps To Save the Ozone Layer.*** Paper 87. Washington, DC: Worldwatch Institute, 1988. 46pp. ISBN 0-916468-88-7.

Like other Worldwatch Papers, this volume presents a well-written, complete introduction to the science, technology, economics, and social aspects of this environmental issue. The paper provides some useful background data on CFC production and use, but it is now somewhat dated.

Silver, Cheryl Simon, ed. ***One Earth, One Future: Our Changing Global Environment.*** Washington, DC: National Academy Press, 1990. 186pp. ISBN 0-309-04141-4.

The book provides a summary of a 1989 conference on global environmental problems sponsored by the National Academy of Sciences. It provides a clearly written summary of a number of important environmental issues.

Stille, Darlene R. ***Ozone Hole.*** Chicago: Childrens Press, 1991. 48pp. ISBN 0-516-01117-0.

One in the "New True Book" series, this book is intended for young readers at the early elementary reading level. The author explains the difference between "good" (stratospheric) and "bad" (tropospheric) ozone, what an "ozone hole" is, how it may have been formed, and what can be done to deal with both forms of ozone.

Tevini, Manfred, ed. ***UV-B Radiation and Ozone Depletion: Effects on Humans, Animals, Plants, Microorganisms, and Materials.*** Boca Raton, FL: Lewis, 1993. 248pp. ISBN 0-8737-1911-5.

This collection of research reports presents one of the most complete summaries of the effects of ultraviolet-B radiation on all forms of living organisms and certain inorganic materials. It is probably the most up-to-date summary in its field.

Titus, James G., ed. ***Effects of Changes in Stratospheric Ozone and Global Climate,*** 4 vols. Washington, DC: Environmental Protection Agency, 1986. 1,140pp. No ISBN.

These volumes provide a summary of an international conference on Health and Environmental Effects of Ozone Modification and Climate Change sponsored by the Environmental Protection

Agency and the United Nations Environment Programme. Volume 1 is an overview that presents a summary of basic information on topics such as Atmospheric Ozone, The Ultraviolet Radiation Environment of the Biosphere, Ozone Perturbations Due to Increases in N_2O, CH_4, and Chlorofluorocarbons, and The Effect of Solar UV-B Radiation on Aquatic Systems. Volume 2 provides more detailed information on the effects of ozone depletion in the area of human health (Nonmelanoma Skin Cancer—UV-B Effects, Effects of UV-B on Infectious Disease, and The Tan of Ultraviolet-B Summer, for example), Aquatic Systems, and Terrestrial Plants. Volumes 3 and 4 deal with climate change issues.

Wayne, Richard P. ***Chemistry of Atmospheres,*** 2d edition. Oxford: Clarendon Press, 1991. 447pp. ISBN 0-19-855574-1.

Wayne's book is perhaps the best general introduction to atmospheric sciences now available. Although the book is intended for students in introductory courses in the atmospheric sciences, it is readily accessible to most general readers. This second edition includes relatively recent information about changes in the Earth's atmosphere brought about by human activities, such as the Antarctic ozone hole.

Weiner, Jonathan. ***The Next One Hundred Years: Shaping the Fate of Our Living Earth.*** New York: Bantam, 1990. 312pp. ISBN 0-553-05744-8.

Weiner has written a very philosophical book about the impact of human activities on the Earth and our environment. He considers a number of specific issues, ozone depletion among them, but examines them from a holistic view.

Wendt, J., J. Kowalski, A. M. Bass, C. Ellis, and M. Patapoff. ***Interagency Comparison of Ultraviolet Photometric Standards for Measuring Ozone Concentrations.*** Washington, DC: U.S. Government Printing Office, 1978. 22pp. No ISBN. Stock number 003-003-02002-8.

This report is a technical bulletin (NBS SP 529) of the National Bureau of Standards comparing a variety of ozone-measuring instruments with each other. The report is quite technical and would be of interest primarily to specialists in the field.

White, James C., ed. ***Global Climate Change Linkages: Acid Rain, Air Quality, and Stratospheric Ozone.*** New York: Elsevier, 1989. 262pp. ISBN 0-444-01515-9.

This book is a compilation of papers presented at a 1988 conference on the relationships of acid rain, air quality, and stratospheric ozone sponsored by the Air Resources Information Clearinghouse, a project of the Center for Environmental Information, Inc. Among the topics included are Linkages between Climate Change and Stratospheric Ozone Depletion, Nitrogen Oxides and Ozone, Policy Linkages, and Environmental Programs during an Era of Deficit Budgets.

World Meteorological Organization. ***Scientific Assessment of Stratospheric Ozone: 1989.*** Geneva: World Meteorological Organization, 1990. 486pp. ISBN 92-807-1255-1.

The 1987 Montreal Protocol called for the convening of four assessment panels. This document is the report of one of those panels, the Scientific Assessment Panel. The four chapters deal with Polar Ozone, Global Trends, Theoretical Predictions, and Halocarbon Ozone Depletion and Global Warming Potentials. The report contains the most recent and most detailed information available at the time. Although much of the text is rather technical, the book contains an executive summary that is readily understandable to the average reader.

Worrest, Robert C., and Martyn M. Caldwell. ***Stratospheric Ozone Reduction, Solar Ultraviolet Radiation and Plant Life.*** Berlin: Springer-Verlag, 1986. 374pp. ISBN 3-540-13875-7.

After outlining some fundamental chemical and physical information about ozone depletion in the opening few chapters, this book focuses on specific and detailed reports on the effects of ultraviolet-B radiation on various living organisms. Included are topics such as Cellular Repair and Assessment of UV-B Radiation Damage, Physiological Responses of Yeast Cells to UV of Different Wavelengths, The Effect of Enhanced Solar UV-B Radiation on Motile Microorganisms, Effects of UV-B Radiation on Growth and Development of Cucumber Seedlings, and Effects of UV-B Radiation on the Growth and Productivity of Field Grown Soybeans.

Zerefos, C. S. ***Atmospheric Ozone: Proceedings of the Quadrennial Ozone Symposium.*** Dordrecht: Reidel Publishing Company, 1985. 842pp. ISBN 90-277-1942-X.

The nine chapters in this very technical book provide an update on the current state of knowledge about atmospheric ozone. The chapters deal with topics such as Chemical-Radiative-Dynamical Model Calculations, Ozone-Climate Interaction, Observations of Relevant Trace Constituents and Their Budgets, Analysis of Ozone Observations, Recent Developments in Observational Techniques, and Interaction of Ozone and Circulation.

Reports

Adequacy of Protection from Sunglasses and Sunscreens. Hearings before the U.S. Senate, Committee on Governmental Affairs, Ad Hoc Subcommittee on Consumer and Environmental Issues, June 5, 1992. Washington, DC: U.S. Government Printing Office, 1993.

EPA Proposed Rulemaking on Chlorofluorocarbons (CFCs) and Its Impact on Small Business. Hearings before the U.S. House of Representatives, Committee on Small Business, Subcommittee on Antitrust and Restraint of Trade Activities Affecting Small Business, July 15, 1981. Washington, DC: U.S. Government Printing Office, 1982.

Global Change Research: Ozone Depletion and Its Impacts. Hearings before the U.S. Senate, Committee on Commerce, Science, and Transportation, November 15, 1991. Washington, DC: U.S. Government Printing Office, 1992.

The Global Change Research Program: Key Scientific Uncertainties. Hearings before the U.S. House of Representatives, Committee on Science, Space, and Technology, Subcommittee on Space, March 30, 1993. Washington, DC: U.S. Government Printing Office, 1993.

Global Climate Change and Stratospheric Ozone Depletion. Hearings before the U.S. Senate, Committee on Environment and Public Works, Subcommittee on Environmental Protection, July 30, 1991. Washington, DC: U.S. Government Printing Office, 1991.

Implications of the Findings of the Expedition To Investigate the Ozone Hole over the Antarctic. Hearings before the U.S. Senate, Committee on Environment and Public Works, Subcommittee on

Environmental Protection and Hazardous Wastes and Toxic Substances, October 27, 1987. Washington, DC: U.S. Government Printing Office, 1988.

New Data on Depletion of the Ozone Layer. Hearing before the U.S. Senate, Committee on Commerce, Science, and Transportation, Subcommittee on Science, Technology, and Space, April 16, 1991. Washington, DC: U.S. Government Printing Office, 1991.

Ozone Depletion, the Greenhouse Effect, and Climate Change. Hearings before the U.S. Senate, Committee on Environment and Public Works, Subcommittee on Environmental Pollution, June 10–11, 1986. Washington, DC: U.S. Government Printing Office, 1986.

Ozone Depletion, the Greenhouse Effect, and Climate Change. Joint hearings before the U.S. Senate, Committee on Environment and Public Works, Subcommittee on Environmental Protection and Hazardous Wastes and Toxic Substances, Part 2, January 28, 1987. Washington, DC: U.S. Government Printing Office, 1987.

Ozone Layer Depletion. Hearings before the U.S. House of Representatives, Committee on Energy and Commerce, Subcommittee on Health and the Environment, March 9, 1987. Washington, DC: U.S. Government Printing Office, 1987.

Ozone Layer Depletion. Hearings before the U.S. House of Representatives, Committee on Energy and Commerce, Subcommittee on Oversight and Investigations, May 15, 1989. Washington, DC: U.S. Government Printing Office, 1990.

Stratospheric Ozone Depletion. Hearings before the U.S. House of Representatives, Committee on Science, Space, and Technology, Subcommittee on Natural Resources, Agriculture Research, and Environment. March 10 and 12, 1987. Washington, DC: U.S. Government Printing Office, 1988.

Stratospheric Ozone Depletion. Hearings before the U.S. House of Representatives, Committee on Energy and Commerce, Subcommittee on Health and the Environment, January 25, 1990. Washington, DC: U.S. Government Printing Office, 1990.

Stratospheric Ozone Depletion. Hearings before the U.S. Senate, Committee on Governmental Affairs, Subcommittee on Consumer and Environmental Affairs, December 17, 1991 and May 15, 1992. Washington, DC: U.S. Government Printing Office, 1994.

Stratospheric Ozone Depletion and Chlorofluorocarbons. Joint Hearing before the U.S. Senate, Committee on Environment and Public Works, Subcommittee on Environmental Protection and Subcommittee on Hazardous Wastes and Toxic Substances, May 12–14, 1987. Washington, DC: U.S. Government Printing Office, 1987.

Stratospheric Ozone: EPA's Safety Assessment of Substitutes for Ozone Depleting Chemicals. Report to the Chairman, Committee on Energy and Commerce, U.S. House of Representatives. Washington, DC: General Accounting Office, 1989.

Nonprint Resources 7

The topic of ozone depletion has not attracted the attention of many filmmakers, video producers, or other publishers of nonprint materials. The resources that are available in this field are summarized in this chapter. All prices are for purchase unless otherwise noted.

Films and Videos

The Amazing Vanishing Ozone
Type: Videocassette
Age: Grades 6 through 12
Length: 30 min.
Cost: $39.95
Date: 1989
Source: PBS Video
1320 Braddock Place
Alexandria, VA 22314-1698
(800) 424-7693; (703) 739-5380
(703) 739-8938 (FAX)

This video focuses on the relationship between CFC emissions and ozone depletion. It explains how human activities may have contributed to the development of this environmental crisis.

Assault on the Ozone Layer

Type: Videocassette
Age: Grade 7 and up
Length: 18 min.
Cost: Purchase $139.00, rental $75.00
Date: 1990
Source: Films for the Humanities and Sciences
Box 2053
Princeton, NJ 08543
(800) 257-5126; (609) 452-1128

Thinning of the ozone layer—over the Antarctic in particular and over the whole globe in general—is the subject of this video. Satellite photographs are used to show how anthropogenic pollutants reach the upper atmosphere and then spread throughout the world. Some possible effects on human health and the natural environment are discussed. The video is part of The Fragile Planet series.

Climate Out of Control

Type: Videocassette
Age: Grade 9 and up
Length: 28 min.
Cost: $149
Date: 1989
Source: Films for the Humanities and Science
Box 2053
Princeton, NJ 08543
(800) 257-5126; (609) 452-1128

Fundamental background information on two major environmental problems, global warming and stratospheric ozone depletion, is presented in this videocassette. After outlining the scientific background of the issues, the program focuses on social, political, and economic consequences of these two global changes. Good use is made of experts in the field, scientists, and government officials.

The Fateful Balance

Type: Videocassette
Age: Grade 7 through adult
Length: 52 min.
Cost: $250
Date: 1992

Source: Landmark Films
3450 Slade Run Drive
Falls Church, VA 22042
(800) 342-4336; (703) 241-2030
(703) 536-9540 (FAX)

Although the focus of this video is on global warming, it may also be useful in understanding some basic ideas about ozone depletion. It shows how modeling is used to predict future trends in the atmosphere and discusses the effects of human activities on macroscopic characteristics of the atmosphere.

A Hole in the Sky
Type: Videocassette
Age: Grade 7 and up
Length: 28 min.
Cost: $250
Date: 1990
Source: Landmark Films
3450 Slade Run Drive
Falls Church, VA 22042
(800) 342-4336; (703) 241-2030
(703) 536-9540 (FAX)

This video provides a general introduction to the topic of ozone depletion. It outlines the chemical reactions by which ozone depletion is thought to occur and it discusses the role of human activities in contributing to this phenomenon. Some long-term consequences of these activities are also considered.

Only One Atmosphere
Type: Videocassette
Age: Grade 10 through adult
Length: 60 min.
Cost: $29.95
Date: 1990
Source: Annenberg/CPB Project
301 E Street, N.W.
Washington, DC 20004
(800) LEARNER; (202) 879-9600
(202) 864-9846 (FAX)

Part of the series Race To Save the Planet, this production describes major changes taking place in the Earth's atmosphere as a result of human activities. The focus is on global warming and depletion of the ozone layer. The video was awarded a 1991 Chris Award for videocassette achievement.

Filmstrips, Slides, and Posters

Blowing Our Cover

Type: Color slides
Age: Grades 8 through 12
Length: 44 slides plus illustrated notes and poster
Cost: £12.60
Date: [no date]
Source: International Center for Conservation Education
Greenfield House, Guiting Power
Cheltenham, Gloucester, England
GL54 5TZ

This collection consists of two sets of slides. The first set of 26 slides deals with stratospheric ozone depletion, covering the origin of the ozone layer, anthropogenic effects on ozone, and consequences of a thinning ozone layer. The second set of 18 slides deals with the topic of global warming. The information in the set is conveyed clearly and in an interesting and appealing fashion.

Our Fragile Atmosphere: Current Issues

Type: Set of two filmstrips
Age: Grades 6 through 9
Length: 55 frames in each filmstrip
Cost: $72
Date: 1991
Source: National Geographic Educational Services
1145 17th Street, NW
Washington, DC 20036
(800) 447-0647; (202) 857-7378

The two filmstrips in this set deal with global warming and ozone depletion. Technically well done, the strips take a fairly standard approach, describing the fundamental scientific background to each problem and showing how human activities have interfered with natural processes. Viewers are encouraged to think about possible approaches for dealing with each problem.

The Ozone Layer

Type: Poster
Age: Middle school
Length: One poster
Cost: £5.40
Date: [no date]
Source: Pictorial Charts Educational Trust
27 Kirchen Road
London, England
W13 0UD

This poster is one in a series dealing with important environmental issues facing the world. Although attractive in layout, a poster can depict only a limited amount of information about stratospheric ozone depletion. The poster is accompanied, however, by a four-page booklet that provides additional background information and teaching ideas about the subject of ozone depletion.

Glossary

This glossary defines a number of terms that are commonly used in discussions of the ozone layer, chlorofluorocarbons, and related topics. A list of important acronyms, chemical symbols, and abbreviations commonly used in these discussions follows the terms and definitions.

aerosol A collection of tiny particles of solid and/or liquid suspended in a gas. Particles in an aerosol range in size from about 0.001 to about 100 microns. Aerosols tend to be very stable and not to settle over long periods of time.

albedo The fraction of solar radiation that strikes a surface, such as clouds, ice, or snow, and is reflected by it.

allotropes Two or more forms of the same chemical element with different physical and chemical properties. For example, diamond and graphite are two allotropes of carbon. Both contain nothing other than carbon in their makeup, but they have strikingly different physical and chemical properties.

anomalous behavior A form of behavior that deviates from that which is expected. The occurrence of the hole in the ozone layer over the Antarctic is an example of an anomalous event.

Antarctic ozone hole A term coined to describe a significant decrease in the concentration of stratospheric ozone over the Antarctic continent.

anthropogenic Resulting from some type of human activity. Chemicals found in the atmosphere may be the result of some natural activity or, as in the case of CFCs, some anthropogenic activity.

atmosphere The envelope of gases that surrounds the Earth.

austral An adjective that refers to the Southern Hemisphere. That is, the austral spring is that period of the year during which spring occurs in the Southern Hemisphere, a period displaced by six months from the corresponding season in the Northern Hemisphere.

backscatter Radiation of an electromagnetic signal by reflection off the ionosphere. For example, when a transmitter sends out a radio signal, most of the signal is propagated in a forward direction. Some, however, bounces off the ionosphere and is reflected back toward the transmitter itself.

best estimate The extrapolation from a physical or mathematical model that appears to provide the most likely description of the future.

blowing agent *See* **neumatogen.**

bonded chlorine Chlorine that is chemically bonded to one or more other elements in a chemical compound. In chlorine monoxide (ClO), for example, chlorine is bonded to oxygen.

bromine A chemical element with atomic number 34 that is used in the synthesis of some important commercial compounds. Among the most important of these are the halons.

Bromine Ozone Depletion Potential (BODP) An indication of the relative effect on the ozone layer of a bromine compound compared to the effect of an equal mass of Halon-1301, the longest lived of all brominated gases. As an example, the BODP of Halon-2402 is about 0.5, meaning that it has about 50 percent the ozone-depleting potential as an equal amount of Halon-1301.

bromocarbons Chemical compounds that contain, at least, carbon and bromine.

"business as usual" A situation in which individuals, industries, and/or nations continue to operate as they have in the past, with no adjustments made in order to deal with some existing or anticipated problem. For example, some observers argue that nations should continue to do "business as usual" until or unless more convincing scientific evidence about ozone depletion can be produced.

CFCs *See* **chlorofluorocarbons.**

Chapman layer A layer of the atmosphere that is energetically favorable to the production and maintenance of ozone. Sydney Chapman first predicted and described the existence of such a layer in 1931. The mathematical expression that he developed to describe the layer is known as a *Chapman function.*

chemical cycle A series of chemical reactions in which one of the products of the final step in the series is also one of the substances needed to begin the first step in the series. Ozone-depleting reactions in the atmosphere, for example, often begin with atomic chlorine, which is also produced as one of the final products of the series of reactions.

chemical feedstock A basic ingredient in the manufacture of other chemicals.

chlorine A chemical element with atomic number 17 that is used in the synthesis of a wide variety of important commercial compounds.

Chlorine Loading Potential (CLP) A measure of the maximum amount of chlorine transported into the stratosphere in comparison with an equal mass of the chlorofluorocarbon CFC-11. For example, the CLP for CFC-113 is 1.11, meaning that CFC-113 releases 111 percent the amount of chlorine to the stratosphere as does an equal mass of CFC-11.

chlorocarbons Chemical compounds that contain, at least, carbon and chlorine.

chlorofluorocarbons (CFCs) Simple organic compounds consisting of carbon, chlorine, and fluorine. They are also sold commercially under the trade name of Freon®.

climate The sum total of weather conditions for a particular area over an extended period of time, normally at least a few decades.

climate model Any numerical simulation of a climate that attempts to describe its behavior. Climate models are used primarily to help scientists make predictions about future climatic conditions.

climatology The study of the climate.

controlled substance A term used in certain international agreements, such as the Montreal Protocol, referring to materials whose production and use are to be limited by those agreements. CFCs are examples of controlled substances mentioned in the Montreal Protocol.

denitrification The chemical process by which nitrogen is removed from a compound.

dobson unit A measure of the concentration of ozone in the atmosphere. One dobson unit (DU) is a layer of ozone 0.01 mm thick at the Earth's surface, i.e., at one atmosphere of pressure. One atmosphere is a standard unit of pressure measurement. It is equal to a pressure of 760 mm of mercury or about 14.7 pounds per square inch.

extrapolation A prediction about some future condition based on what is known about past trends and the present state of affairs. As an example, atmospheric scientists try to make extrapolations about the concentration of stratospheric ozone in the twenty-first century based on what they know about current conditions and about factors that might change those conditions in the future.

false color images Ozone concentration graphs that have been given artificial colors in order to represent more clearly the differences obtained in a particular observation.

feedback mechanism Any series of changes in which the final step in the series results in an outcome that influences the first step in the series. A *positive feedback mechanism* is one in which the series of changes is amplified in later repetitions of the cycle, while a *negative feedback mechanism* is one in which the series is mitigated over time. An example of a positive feedback mechanism is the series of chemical reactions by which free chlorine atoms destroy ozone molecules. See pages 8–9.

fluorocarbons A group of organic compounds containing carbon and fluorine atoms.

free radicals Particularly active forms of chemical elements or compounds. Free radicals have been implicated in the chemical processes by which ozone is destroyed in the stratosphere.

Freon® The trade name for a group of CFCs produced by the Du Pont Corporation.

Gaia The Greek mother-goddess. The British scientist James Lovelock appropriated the term to describe his theory that human life exerts an important influence on the planet, altering land, air, and sea in such a way as to make the planet more hospitable to human life. In this sense, the planet is seen almost as a living entity itself. The Gaia hypothesis has been the subject of intense scientific debate for more than 20 years.

global radiation The total solar radiation reaching the Earth's surface from the sky, including both direct radiation and radiation scattered from the sky and clouds.

global warming The term used to express the apparent increase in average annual global temperatures that has been taking place over the past century or less. Gases involved in ozone depletion are, in many cases, also implicated in global warming.

global warming potential (GWP) A mathematical measure of the tendency of any given substance to contribute to global warming. For the purposes of calculating this value for ozone-depleting substances, all materials are compared to CFC-11, which is assigned a value of unity (1).

greenhouse effect A natural phenomenon in which gases in the atmosphere, especially carbon dioxide, capture heat radiated from the Earth's surface. The result of the greenhouse effect is that the Earth is significantly warmer than it would be without the presence of these gases. The increased concentration of carbon dioxide in the atmosphere as a result of human activities is now thought to be increasing the natural effect of the greenhouse phenomenon.

greenhouse gas Any gas that does not absorb solar radiation but does absorb long-wavelength radiation reflected from the Earth's surface. Although carbon dioxide is probably the most important naturally occurring greenhouse gas, CFCs and other ozone-depleting gases are also greenhouse gases.

halocarbon A chemical compound that contains carbon and one or more of the halogen elements (fluorine, chlorine, bromine, and iodine).

halogen The name given to a family of chemical elements. Members of the family are fluorine, chlorine, bromine, iodine, and astatine (a rare, radioactive member of the family).

halon The commercial name for a group of fluorocarbons.

HCFCs *See* **hydrochlorofluorocarbons.**

heterogeneous chemical reactions Chemical reactions that involve both liquid and gaseous phases.

HFCs *See* **hydrofluorocarbons.**

homogeneous chemical reactions Chemical reactions in which all substances taking part are in the same phase (solid, liquid, or gas).

hydrochlorofluorocarbons (HCFCs) Organic compounds that contain carbon, hydrogen, chlorine, and fluorine. HCFCs have been used as substitutes for CFCs in many commercial and industrial applications.

hydrofluorocarbons (HFCs) Organic compounds that contain carbon, hydrogen, and fluorine. Some HFCs may be acceptable substitutes for CFCs in certain commercial and industrial applications.

hydroxyl radical A highly reactive chemical substance made of an atom of hydrogen and an atom of oxygen joined to each other.

industrial rationalization A procedure that allows one nation or region to transfer all of part of the production of a controlled substance (such as a CFC) to another nation or region. Industrial rationalization is a method for maintaining production or use of an ozone depleter at a certain given level worldwide, while allowing one nation to produce or use more than would normally be allowed by an agreement.

in situ A term that means "in place." *In situ* measurements of ozone concentrations are somewhat difficult to make since they require that some means be found by which scientists can actually travel to the stratosphere and measure ozone concentrations there. As a consequence, many ozone measurements are made by *remote* techniques, as from satellites or Earth-based stations.

insolation The amount of solar radiation that falls on a unit area of horizontal surface in a given period of time.

irradiance A measure of the total amount of radiant power that falls on a unit of area. Irradiance is also called *radiant flux density.*

kelvin (K) A scale commonly used by scientists to measure temperatures, especially very low temperatures. A temperature of 0 K (zero degrees kelvin) is equal to about –273°C or –459°F.

lidar An acronym for light detection and ranging system. Lidar is an instrument that converts radiation in the form of visible light or infrared radiation into a laser beam that can then be used to measure wind speed and direction. Lidar is used in the study of wind-borne aerosols.

melanoma A type of skin cancer that is often fatal. Although the cause(s) of melanoma are not fully understood, ultraviolet radiation is widely regarded as a major cause of the disease. Melanomas are often distinguished as malignant and nonmalignant.

millibar (mb) A unit used to measure pressure; one-thousandth of a bar. A millibar is equal to 2.09 pounds per square foot in the English system.

mixing ratio A method for expressing concentration commonly used by atmospheric scientists. The term is identical to the measure widely used among chemists known as *mole fraction.*

model Most commonly in ozone studies, a set of mathematical equations developed by researchers designed to simulate the properties and/or behavior of some natural phenomenon. Ozone models, for example, are designed to predict future changes in stratospheric ozone as a result of changes in the release of ozone-depleting chemicals by human activities. Depending on the degree of complexity involved, most ozone models are designated as 1-D (simpler), 2-D (more complex), or 3-D (most complex) models.

Montreal Protocol An international agreement signed on 16 September 1987 to limit and eventually reduce the quantities of CFCs and other ozone-depleting chemicals released into the atmosphere.

nanometer A unit of linear measurement equivalent to one-billionth (10^{-9}) of a meter.

Nicolet diagrams Pictorial representations that show how various chemical species are related to each other in a region of the atmosphere. A Nicolet diagram consists of at least two ovals containing the names of two different chemical species and sets of arrows showing how one species is converted into another.

nitrous oxide A chemical compound with the formula N_2O, found in trace amounts in the atmosphere. Nitrous oxide is also produced as a result of anthropogenic activities and is believed to have a significant effect on atmospheric properties.

ozone An allotrope of oxygen that consists of three atoms to the molecule (O_3) rather than the more common two atoms to the molecule (O_2) found in atmospheric diatomic oxygen. The thin layer of ozone in the stratosphere absorbs ultraviolet radiation, thereby protecting plants and animals on the Earth's surface from damage.

Ozone Depletion Potential (ODP) An indication of the relative effect on the ozone layer of any given chemical compared to the effect of an equal mass of the chlorofluorocarbon known as CFC-11. As an example, the ODP of CFC-12 is about 0.9, meaning that it has about 90 percent the ozone-depleting potential as an equal amount of CFC-11.

ozone hole A significant reduction in the concentration of ozone in the stratosphere. The term was originally and is most commonly used with reference to the reduction of ozone over the Antarctic continent. *See also* **Antarctic ozone hole.**

ozonesondes A term with two meanings: (1) Instruments used in the measurement of ozone concentrations in the atmosphere; various types exist, including balloon, electrochemical concentration cell (ECC), and rocket ozonesondes; (2) vertical profiles of ozone concentrations in the atmosphere.

parts per million, parts per billion, parts per trillion (ppm, ppb, ppt) Units to express the concentration of materials that appear in very low amounts. One part per million (1 ppm), for example, is the equivalent of a concentration of 0.0001%. The concentration of many gases in the atmosphere can be expressed in these units. The addition of the letters m or v indicate that the concentration is expressed in mass units or volume units, respectively.

perturbation Some change that takes place in a system because of the presence of a particular substance or event. For example, the release of chlorine compounds into the stratosphere has brought about a perturbation of the ozone layer found there.

photochemistry The study of chemical changes that take place as the result of some form of radiation energy.

photodissociation The process by which radiation energy causes a molecule to break apart into two or more pieces. Photodissociation is also known as photolysis.

photolysis *See* **photodissociation.**

phytoplankton Simple, one-celled organisms that exist in huge numbers in the oceans. The most common examples of phytoplankton are algae and diatoms.

pneumatogen A gas used for inflation. Air, nitrogen, carbon dioxide, and CFCs are common pneumatogens. In the field of polymer foam technology, a pneumatogen is used to produce a material with low density. Pneumatogens are also known as blowing agents.

polar ice caps Vast regions of deep ice and snow that cover the areas surrounding the South and North Poles.

polar stratospheric clouds (PSCs) Very high clouds that form over the polar regions during the coldest part of the year. Because the Antarctic is colder than the Arctic, PSCs are much more common over the South Pole than over the North Pole. Most scientists agree that PSCs have some role in the depletion of stratospheric ozone.

polar vortex A strong belt of winds encircling the South Pole at latitudes of about 60°S to 70°S. A similar phenomenon exists in the region around the North Pole during the coldest months of the year.

precursor A chemical substance from which some other substance is formed. For example, chlorofluorocarbons are precursors of chlorine in the stratosphere. Chlorine itself is then a precursor of chlorine monoxide.

protocol Most commonly, an early draft from which some form of international agreement is developed. The Montreal Protocol is, in this regard, somewhat misnamed since it was a final treaty itself rather than a preliminary document leading to a more complete treaty. The word *protocol* is a also commonly used by scientists to describe the outline or plan that they will follow in a particular research project.

quasi-biennial oscillation (QBO) A natural oscillation of stratospheric winds over the equator. QBO is known to be responsible for at least some portion of ozone thinning that occurs over the Antarctic each austral spring.

radiative budget or **radiation budget** A system for accounting for all of the radiant energy that enters and leaves the Earth's atmosphere. The radiative budget is similar to a financial budget in which the income and expenditure of all monies are distributed to various accounts. The radiative balance is also known as the *radiative balance* or *radiation balance.*

radiosonde A measuring device carried into the atmosphere by balloons. It is used to measure atmospheric pressure, temperature, and humidity.

rate constant A mathematical measure of the rate at which a certain chemical reaction takes place.

residence time The time during which some substance remains in a particular reservoir. In ozone research, for example, it is essential to know how long a particular ozone-depleting chemical can be expected to remain in the atmosphere.

retrofit Upgrading of existing equipment or facilities, often in order to meet new regulations.

Robertson-Berger meter One of the instruments commonly used to measure ozone levels in the stratosphere.

SBUV An acronym for the solar backscatter ultraviolet instrument used aboard satellites in order to measure ozone concentrations in the stratosphere.

scenario A plan, design, or prediction regarding some future outcome. For example, researchers attempt to find out what scenarios for the ozone layer would occur at some point in the future if CFC production stays the same, is increased, or is decreased by some given amount.

self-correcting mechanism Any process that operates in such a way as to reduce or eliminate some trend in the process itself. For example, loss of ozone in the stratosphere contains a (limited) self-correcting mechanism in the sense that the greater the number of ozone molecules that are destroyed, the greater the number of oxygen molecules that are created, and, therefore, the greater the chance that new ozone molecules will be created. Any self-correcting mechanism, including this one, can be overwhelmed, however, by the introduction of foreign materials that destroy the mechanism's ability to continue.

signal-to-noise ratio A numerical measure of the significance of changes that are detected in some natural process. For example, the concentration of ozone in the stratosphere changes naturally from hour-to-hour, day-to-day, and month-to-month depending on a number of factors. Those changes constitute a "background noise" against which more profound changes (such as those caused by CFCs) must be compared. Scientists must be able to show that the "profound changes"—the signal—are really distinct from the background noise in order to make their case for change.

simulation An attempt to imitate or reproduce some physical phenomenon using mathematical models, physical models, or both. For example, researchers attempt to simulate the stratosphere over the Antarctic 20 years from now by writing computer programs that can predict changes in ozone concentration that would take place as a the result of changes in CFC emissions.

spectrophotometer An instrument used to measure the amount of radiation at various wavelengths.

steady state A condition in which a system is no longer undergoing any changes or variations.

sulfate aerosols Aerosols that consist of materials produced by the interaction of sulfur dioxide (SO_2) and sulfur trioxide (SO_3) with other substances in the atmosphere.

stratopause The upper boundary of the stratosphere; the region between the stratosphere and the ionosphere.

stratosphere The portion of the atmosphere between the upper boundary of the troposphere, a distance of about 15 kilometers (9 miles) above the Earth's surface, and the lower boundary of the ionosphere, a distance of about 50 kilometers (30 miles) above the Earth's surface.

Sun Protection Factor (SPF) A number assigned to sunscreen products that refers to the material's ability to protect skin from burning by UVB radiation compared to the skin's natural protection. Skin treated with SPF 15, for example, will be exposed to the same amount of UVB radiation in 15 minutes as will untreated skin in one minute.

synoptic-scale event An event that has been observed and measured at a large number of observing stations located at widespread locations around the Earth.

technological fix An attempt to solve some social, environmental, or other problem by using science or technology rather than politics, social forces, economics, or other approaches.

TOMS An acronym used for the total ozone mapping spectrometer, an instrument carried into space on a variety of satellites and used to measure ozone concentrations in the stratosphere.

total column ozone The total amount of ozone in a column of the atmosphere above a certain point on the Earth's surface. Measurements of total column ozone allow scientists to detect overall changes in ozone concentration from one period of time to another.

trace gas A gas that appears in the atmosphere in very low concentrations.

tracer A substance whose location in or movement through a system can be followed by some external means, such as by a radiation detection device.

transitional substances A term used in the London Revisions to the Montreal Protocol referring to substances listed in Annex C of the protocol, the hydrochlorofluorocarbons. The term is used because these compounds appear to be less destructive of the ozone layer than are chlorofluorocarbons. Still, they do appear to have some deleterious effects and will, therefore, need to be replaced themselves at some time in the future.

trend analysis A mathematical attempt to detect significant changes that have taken place in some physical phenomenon over a period of time.

tropopause The boundary between the troposphere and stratosphere.

troposphere The layer of the atmosphere that lies closest to the Earth's surface. Its upper boundary lies at an average distance of about 15 kilometers (9 miles) above the Earth.

ultraviolet radiation A form of electromagnetic radiation with wavelengths ranging between about 4 and 400 nanometers. Ultraviolet radiation is often subdivided into two categories, ultraviolet-A (UVA) and ultraviolet-B (UVB) radiation. The former term refers to ultraviolet radiation with wavelengths in the range 320 to 400 nanometers, while the latter refers to radiation with wavelengths in the range 240 to 320 nanometers. Of the two forms of ultraviolet radiation, UVB is more harmful to plants and animals (including humans) than is UVA.

umkehr effect Variations in the intensities of sunlight scattered as the sun approaches the horizon. The umkehr effect is caused by the ozone layer and is, therefore, an important measure of the amount of ozone present in that layer.

uncertainty The extent to which some measurement is not known. All scientific measurements are subject to some uncertainty because of inadequate measuring tools, inaccessibility of the object or event being measured, human error, and other factors. The uncertainty of a specific measurement is usually indicated mathematically by means of a plus-or-minus (±) symbol attached to the measurement. For example, a measurement of 3.4 ± 0.7 means that the most likely value for the measurement (3.4) could be incorrect by as much as 0.7 in either direction. That is, the measurement could actually be anywhere from 2.7 (3.4 –0.7) to 4.1 (3.4 + 0.7).

vertical profiles Variations in some measure, such as atmospheric pressure, for a continuously varying section of the atmosphere. Ozone vertical profiles show, for example, the variations in ozone concentrations in a vertical column of the atmosphere above a particular point on the Earth.

vortex A movement of air consisting of a very cold core surrounded by rapidly moving winds. Vortices form over both the Antarctic and Arctic, but are much more common above the former.

Acronyms and Abbreviations

AAOE	Airborne Antarctic Ozone Experiment
AASE	Airborne Arctic Stratospheric Expedition
AFEAS	Alternative Fluorocarbon Environmental Acceptability Study
AFGL	Air Force Geophysical Laboratory
AGU	American Geophysical Union
ALE/GAGE	Atmospheric Lifetime Experiment/Global Atmospheric Gases Experiment
AMAP	Association for Meteorology and Atmospheric Physics
ARC	Ames Research Center (NASA)
ATMOS	Atmospheric Trace Molecule Spectroscopy
AVHRR	Advanced Very High Resolution Radiometer
BAU	Business as Usual
BLP	Bromine Loading Potential
BUV	Backscatter Ultraviolet Spectrometer
CFC	Chlorofluorocarbon. *See also glossary.*
CFM	Chlorofluoromethane
CGC	Committee on Global Change
CHEOPS	Chemistry of Ozone in the Polar Stratosphere

CLP	Chlorine Loading Potential. *See also glossary.*
CMA	Chemical Manufacturers Association
CMRN	Cooperative Meteorological Rocketsonde Network
COSPAR	Committee on Space Research
CPOZ	Compressed Profile Ozone
DU	Dobson Unit
EGA	Emissivity Growth Approximation
EI	Emission Index
EMR	Electromagnetic Radiation
ENSO	El Niño-Southern Oscillation
EOS	Earth Observing System
EPA	Environment Protection Agency
ERBS	Earth Radiation Budget Satellite
ERL	Environmental Research Laboratory (NOAA)
ESA	European Space Agency
GAGE	Global Atmospheric Gas Experiment
GARP	Global Atmospheric Research Program
GFDL	Geophysical Fluid Dynamics Laboratory
GHRS	Goddard High Resolution Spectrograph
GMT	Greenwich Mean Time
GSFC	Goddard Space Flight Center (NASA)
GWP	Global Warming Potential. *See also glossary.*
HCFC	Hydrochlorofluorocarbon. *See also glossary.*
HFC	Hydrofluorocarbon. *See also glossary.*
HIRS	High Resolution Infrared Radiation Sounder
ICSU	International Council of Scientific Unions
IFC	Inflight Calibration
IFOV	Instrument Field of View
IOC	International Ozone Commission
IPCC	Intergovernmental Panel on Climate Change
IR	Infrared
JPL	Jet Propulsion Laboratory
LIMS	Limb Infrared Monitor of the Stratosphere
LRIR	Limb Radiance Inversion Radiometer
LTE	Local Thermodynamic Equilibrium
MAP	Middle Atmosphere Program
MLS	Mid-Latitude Summer
MSU	Microwave Sounding Unit
NASA	National Aeronautics and Space Administration
NBS	*See* **NIST**
NCAR	National Center for Atmospheric Research
NDSC	Network for the Detection of Stratospheric Change
NESDIS	National Environmental Satellite Data and Information Service
NIR	Near Infrared
NIST	National Institute for Standards and Technology; formerly National Bureau of Standards (NBS)

NMHCs	Nonmethane Hydrocarbons
NOAA	National Oceanic and Atmospheric Administration
NOAA/AL	National Oceanic and Atmospheric Administration/Aeronomy Laboratory
NOZE	National Ozone Expedition
NRC	National Research Council
NSSDC	National Space Science Data Center
NWS	National Weather Service
ODP	Ozone Depletion Potential. *See also glossary.*
ODW	Ozone Date for the World
OEDC	Organization for Economic Cooperation and Development
OGO	Orbiting Geophysical Observatory
OTP	Ozone Trends Panel
PAN	Peroxyacetyl Nitrate
PMR	Pressure Modulated Radiometer
PMT	Photomultiplier Tube
PPN	Peroxypropionyl Nitrate
PSC	Polar Stratospheric Clouds. *See also glossary.*
PV	Potential Vorticity
QBO	Quasi-Biennial Oscillation. *See also glossary.*
RAF	Radiative Amplification Factor
RAOB	Rawinsonde Observation
RB	Robertson-Berger Network
ROCOZ	Rocket Ozonesonde
SAGE	Stratospheric Aerosol and Gas Experiment
SAM II	Stratospheric Aerosol Measurement
SBUV	Solar Backscatter Ultraviolet. *See also glossary.*
SCOSTEP	Scientific Committee on Solar Terrestrial Physics
SCR	Selective Chopper Radiometer
SIRS	Stratospheric Infrared Interferometer Spectrometer
SME	Solar Mesosphere Explorer
SMM	Solar Maximum Mission
SOI	Southern Oscillation Index
SPEs	Solar Proton Events
SST	Supersonic Transport
SSU	Stratosphere Sounding Unit
SZAs	Solar Zenith Angles
THIR	Temperature Humidity Infrared Radiometer
TIROS	Television and Infrared Observation Satellite
TOMS	Total Ozone Mapping Spectrometer. *See also glossary.*
UADP	Upper Atmosphere Data Program
UARS	Upper Atmosphere Research Satellite
UNEP	United Nations Environment Programme
UV	Ultraviolet
UVB or **UV-B**	Ultraviolet-B
UVS	Ultraviolet Spectrometer

UVSP Ultraviolet Spectrometer and Polarimeter
VTPR Vertical Temperature Profile Radiometer
WMO World Meteorological Organization
WODC World Ozone Data Center

Formulas and Nomenclature of Important Chemical Species

O	Atomic (monotomic) oxygen
O_2	Molecular (diatomic) oxygen
O_3	Ozone
O_x	Odd oxygen
N_2	Molecular nitrogen
N_2O	Nitrous oxide
NO	Nitric oxide
NO_3	Nitrogen trioxide
NO_x	Oxides of nitrogen
NO_y	Odd nitrogen
HNO_2	Nitrous acid
HNO_3	Nitric acid
HNO_4	Peroxynitric acid
NH_3	Ammonia
H_2O	Water or water vapor
H_2O_2	Hydrogen peroxide
OH	Hydroxyl radical
HO_2	Hydroperoxyl radical
HC	Hydrocarbon
CO	Carbon monoxide
CO_2	Carbon dioxide
CS_2	Carbon disulfide
COS	Carbonyl sulfide
CH_4	Methane
C_2H_6	Ethane
C_2H_4	Ethene; ethylene
C_2H_2	Ethyne; acetylene
CH_2O	Formaldehyde
CH_3CHO	Acetaldehyde
CH_3O_2H	Methyl hydroperoxide
C_2Cl_4	Tetrachloroethylene
CH_3Cl	Methyl chloride; Monochloromethane

CH_2Cl_2	Dichloromethane
$CHCl_3$	Chloroform; Trichloromethane
CCl_4	Carbon tetrachloride; Tetrachloromethane
CH_3CCl_3	Methyl chloroform
CF_4	Tetrafluoromethane
C_2F_6	Hexafluoroethane
CCl_3F	Trichlorofluoromethane (CFC-11)
CCl_2F_2	Dichlorodifluoromethane (CFC-12)
$CClF_3$	Chlorotrifluoromethane (CFC-13)
CCl_2FCClF_2	Trichlorotrifluoroethane (CFC-113)
$CClF_2CClF_2$	Dichlorotetrafluoroethane (CFC-114)
$CClF_2CF_3$	Chloropentafluoroethane (CFC-115)
$CHCl_2F$	Dichlorofluoromethane (HCFC-21)
$CHClF_2$	Chlorodifluoromethane (HCFC-22)
$CBrClF_2$	Bromochlorodifluoromethane (Halon-1211)
$CBrF_3$	Bromotrifluoromethane (Halon-1301)
SO_2	Sulfur dioxide
H_2SO_4	Sulfuric acid
HF	Hydrogen fluoride
HCl	Hydrogen chloride
HOCl	Hypochlorous acid
Cl	Atomic chlorine
Cl_2	Molecular chlorine
ClO	Chlorine monoxide
$ClONO_2$	Chlorine nitrate
Cl_x	Odd chlorine
Br	Atomic bromine
BrO	Bromine monoxide
Br_x	Odd bromine
CH_3Br	Methyl bromide

Index